Engineering Your Academic Career

John L. Junkins
Texas A&M University

Brief Description: Professor Junkins draws upon his four decades as a professor to provide candid, to-the-point mentoring advice for aspiring academics. Dr. Junkins has been successful in the four critical aspects of being a professor:

(i) *Mentoring* (directing almost four dozen PhD students to completion, giving rise to three descendent generations of PhDs).

(ii) *Teaching* (having won numerous teaching awards and impacting a generation of students).

(iii) *Scholarship* (having published six books, over four hundred publications, and several patents).

(iv) *Sponsored research* (having won support and served as Principal Investigator on over 80 projects funded by over 20 agencies, with a total budget of about $40M).

Dr. Junkins shares insights on how to succeed with a smile while balancing the above four activities and other dimensions of academic life with the rest of your life. The insights are drawn from Junkins' personal teaching and research, but perhaps more importantly, from successful mentoring collaborations with a number of young professors while they navigated the challenges of modern university life.

ISBN 978-1-105-31585-5

LULU.COM   2012

Preface

This book is designed to help you young and mid-career professors navigate the first two decades of your career. I hope my more senior colleagues will also find things to ponder with regard to enhancing the modern academic culture, so that it is more conducive to developing the next generation of outstanding professors. I offer no-nonsense advice drawn from my experience with emphasis on helping you budget that precious fluid of life (your time) while you find happiness as a teacher, mentor, and researcher. This is not a cookbook to control your calendar or activities, but rather some straight talk to help you develop an internal compass that points to a good path leading to success and happiness. While much of the information provided is opinionated and anecdotal, I do also give data that shows in a statistical sense "what it takes" to succeed in a top engineering program. I will also provide insight to help you judge what such statistics really mean with regard to genuine quality.

Modern engineering academic life is very busy. Those who succeed will find their lives both challenging and wonderfully rewarding. On the other hand, I have observed that a significant fraction of my young colleagues experience occasional high levels of frustration during the early years of their careers. I realized some time ago that one of the fundamental problems facing young professors was that their education and experience prior to entering a tenure-track position had not sufficiently prepared them for all of the challenges they face. They must efficiently learn to be on the other side of the desk in teaching and mentoring, while simultaneously learning the "business" aspects of acquiring and conducting sponsored research. This latter aspect (grantsmanship) is "un-natural" to some individuals. As explained herein, easy or hard, attracting significant research sponsorship is necessary, given the structure of the modern university. Concrete suggestions are made that I have observed to vastly accelerate writing of successful research proposals.

This book also deals with effective teaching, mentoring of graduate students, as well as professional and university service.

As is likely obvious to you, we live in an environment where everyone seems obsessed with being (or at least, appearing) to be highly meritorious. A meritocracy is the natural end state of a university, but the stimuli sent to young professors can be confusing.

In some academic cultures, there is hesitation to "tell it like it is," with regard to open discussion of "target levels" of various performance metrics (measuring quality of teaching, research, publications, citations, research dollar volume needed to succeed, etc.). I will address this deficit by describing success from a statistical point of view, in top programs across the nation. However, I *do not* subscribe to the notion that achieving a statistical prescription guarantees anything with regard to your professional and academic health and success, and more importantly, satisfying yourself. Measuring your performance with purely statistical approach is naïve, and in lieu of offering numerical recipes, I communicate clearly the things I think you need to understand and act upon to succeed and find happiness.

In the course of this presentation, we will use history of stellar academics to help gain insights needed to fully understand the culture of modern engineering. Together, we will revisit the remarkable careers of two giants of our profession: Willard Gibbs and Theodore von Kármán, on whose shoulders we all presently stand. We will also consider two anonymous modern-day professors X and Y, to help focus your ambition with regard to achieving technical depth versus breath, and also the degree to which you wish to emphasize service and leadership roles in your profession.

John L. Junkins
College Station, Texas
January, 2012

Chapter One

Introduction

Probably because my career as a professor now rounds off to four decades, during recent years I am frequently asked questions such as, "what does it take" to:

- Qualify for promotion and tenure as an Associate Professor?
- Achieve promotion to Full Professor?
- Make a significant impact on the literature?
- Become a leader in the field?
- Advance the state of engineering practice?
- Become nationally recognized for my contributions?

I usually begin by telling my younger colleagues to cultivate a genuine passion for quality research, teaching, and service, and through investment of hard work, the future (and the coupled answer to the above questions) will mostly take care of itself. Of course this platitude of an answer, while true, is not useful in setting goals and priorities. In this text, I give a more responsive and "actionable" set of answers and provide constructive suggestions that should help you budget that precious fluid of life, your time. I also do my best to provide a context that will assist you in answering a more fundamental question that may be on your mind: "Should I pursue a career as a professor?"

To set the stage, I begin by stating what I feel is the most fundamental thing you need to do is "get your mind right." [1]

Getting your mind right involves:

- Deciding if you are sufficiently interested in an academic career to pay your dues.

- Investing the time necessary to build your technical foundation with sufficient depth and breadth to develop a first-rate academic career.[2]

- Buying into the notion: "the most important responsibility of a university professor is to nurture the academic development of students".[3]

- Developing the self-discipline and cultivating the joy of learning needed to expand your spheres of competence throughout your career.

- Making the commitment to become a career scholar and excellent teacher/mentor.

- Approaching the challenge as a "labor of love marathon", rather than a sprint.

---

[1] From the 1967 movie Cool Hand Luke. "Get your mind right, Luke" - lines spoken by the redneck prison warden Boss Paul Hunnicut (played by Luke Askew) to the rookie prisoner Lucas Jackson (played by Paul Newman). Hunnicut was "mentoring" Jackson to accelerate his adjustment to life in prison. "Get your mind right, ..." is the second most famous quotation from Boss Hunnicut, who summarized his inability to break Lucus Jackson's spirit, following a brutal beating, when he spoke with a thick southern accent the immortal words: "What we have here is a failure to communicate!" This reference is certainly not intended to suggest that mentoring of young professors has anything to do with adapting to prison life on the "chain gang!"

[2] The most fundamental necessary condition is your technical preparation. If you judge that your PhD preparation was not sufficiently rigorous, or if you need a little more seasoning, consider a post-doctoral research appointment prior to entering academe on a tenure-track appointment.

[3] If you believe the most important function of a university professor is optimizing your personal research progress, then you should likely pursue your career elsewhere. Research and personal scholarship are vitally important, of course, but must be time-shared in a way that optimizes the teaching/mentoring of stellar students. If you enjoy a long and successful academic career, it is virtually certain that your students will prove to be your most important legacy.

- Cultivating your ability to evaluate possible research problems critically.

- Being conservative when obligating your time (more on this later), don't commit to pursue a research direction unless you feel it is a good fit.

- Recognizing that your ability to get along with colleagues will be at least as important as your overall professional competence.

- Being a positive force on your faculty; the esprit d 'corps, especially at the departmental level, is a fundamental element in the quality of academic life; department collegiality can be badly damaged by a few "nabobs of negativity[4]".

So, I recommend that you commit to cultivating your ability to find common ground with fellow faculty members and to do more than your share to improve the professional climate.

My four decade academic career has been spent on the faculties of three public universities. While this experience informs my advice, much of what I have to say will also map easily into meaningful advice if you serve on the faculty of a top private

---

[4] A linguistic contribution of Richard Nixon's combative Vice President Spiro Agnew was his circa 1970 witty barb: "In the United States today, we have more than our share of the *nattering nabobs of negativism*, and they have formed their own 4H Club: The *hopeless, hysterical hypochondriacs of history*." Agnew was demeaning the "talking head critics" in the press and his political opponents who were raising various questions about the Nixon Administration that ultimately culminated in Nixon's impeachment and Agnew being sent to prison for accepting bribes and falsifying his tax returns. A lesson from this dark chapter in US history is that negativity is occasionally justified, but let it always be in response to exceptional wrongs rather than just being a defining aspect of one's personality! Clearly having a "negative set-point" for your interactions with colleagues will make you less fun to be around. Having a generally positive mode of collegial interaction will increase your impact and effectiveness for those unusual circumstances where you challenge colleagues and to address a serious problem at hand.

institution. I make this inference based on interaction with colleagues across the spectrum of public and private institutions and also from statistical analysis of performance metrics. Perhaps the largest differences between a strong public and a strong private university lies in the demographics and size of the student bodies during the first two years of undergraduate study.

Over 20 of my technical offspring have joined the professorate and as a consequence, I have three generations of PhD offspring. In addition to drawing on my first-hand, and second-hand anecdotal experiences, I have solicited statistical information from informal surveys of colleagues at four institutions (MIT, Purdue University, Texas A&M University, and *the* University of Texas at Austin[5]). While the statistical information is useful to numerically describe the distribution of metrics associated with achieving promotion and tenure, the most important aspects of this book are heuristic and philosophical insights. If these insights are acted upon, they can be decisive in helping you achieve success and happiness in academia.

I offer now a few remarks with respect to the focus of this book relative to related literature. There exist several books that address the critical issue of how to teach engineering effectively, with detailed advice on how to prepare lectures, exams, and interact with students. I believe the most comprehensive of these is the 1993 book *Teaching Engineering* by Wankat, and Oreovicz. [6] With regard to career advice, I mention there exists another interesting book *Advice to Rocket Scientists* by Jim

---

[5] I can't resist a good-natured jab at my sister institution in Austin. Whenever I see a University name with an italic *the* in front of the institution's name, I am reminded of a 1977 seminar I gave at *the* University of Texas. I was then a young professor at *the* University of Virginia, where our football team was mid-way on a 30 game losing streak. Our athletic director had the audacity to schedule a game at *the* University of Texas during the Heisman Trophy senior season of Earl Campbell. At my seminar, about 2 weeks after a 68 – 0 shellacking by *the* University of Texas, I was introduced by my friend, UT Professor Byron Tapley, who said with a satisfied smirk "*the* University of Virginia was just here 2 weeks ago, and let me tell you, I was impressed by *the* University of Virginia!". Following a rowdy laugh from my audience, Bryon continued his friendly roast, he said "Oh I don't mean your football team, I mean your cheerleaders. At halftime, trailing by 35 – 0, your cheerleaders carried around a sign in front of 80,000 Texas fans that read: We were *the* University of Virginia before Texas was *a* State!"

[6] Wankat, P. C., and Oreovicz, F. S., *Teaching Engineering*, McGraw-Hill, ISBN0-07-068154-6, New York, 1993.

Longuski.[7] Longuski gives insightful career advice for engineers pursuing mainly non-academic careers. Most of Longuski's advice, however, is relevant to engineering professionals broadly, not just aerospace professionals in industry and academia.

While the present book overlaps some with *Teaching Engineering*, a different approach is adopted. Instead of detailed advice on the mechanics of effective teaching, here I choose instead to focus mainly on *mentoring advice* for succeeding with a smile in your academic career. The target audience is specifically addressed to aspiring professors, in contrast to the broader aims of *Advice for Rocket Scientists*.

I have written this book in an informal, first person, "big brother" conversational style that I hope will be found easily understood and applied to your career. As engineers, it is easy to adopt the blinders-on notion that developing over-powering technical competence will guarantee success as a professor. However, my observation is that strong technical competence is just one of the necessary conditions for success in the modern university. At the heart of being happy and achieving my definition of success is "getting your mind right" as I mentioned above.

Upon digesting this book you will have the insights to answer such critical questions as:

- How is excellence judged in the tenure and promotion process?

- How can I achieve and maintain a valid perspective on the critical research problems in my discipline?

- How should I approach the mentoring of graduate students?

---

[7] Longuski, J., *Advice to Rocket Scientists: A Career Survival Guide for Scientists and Engineers*, American Institute of Aeronautics and Astronautics, ISBN 1-56347-655-X, Washington, DC, 2004.

- Is it teaching *and* research, or teaching *versus* research?

- How can I develop a research program that embodies, to a maximum extent possible, mentorship of students and development of their knowledge?

- How can I develop a collegial network with peers in governmental and industrial laboratories, leading to research sponsorship, without feeling like a traveling salesman?

- How do I balance my involvement in basic engineering science research and more applications-oriented engineering research?

- How can I improve my chances of writing winning proposals?

- What should I do when journal papers and/or research proposals get rejected?

- Why is research funding considered so important?

- Is the judgment of my research excellence really just a numbers game?

- How do I tactfully say no to unreasonable requests from senior colleagues?

Through statistical analysis and descriptions of various measures of scholarship, teaching, mentoring and research sponsorship associated with achieving tenure and promotions at top institutions, you will achieve an informed qualitative appreciation of what the expectations are in top institutions. More importantly, you will develop insights on how to think about these data and how they relate to genuine excellence. Finally, I give you some rules of thumb to help you budget your effort and time in a way that will position you for success and happiness.

To set the stage for the remaining chapters of this book, I direct your attention to Figure 1 where I depict an "Alice in Wonderland" idealization of a successful academic career. This career trajectory might correlate qualitatively to an actual career (actually, it is not a bad representation of my own life), but it sweeps under the carpet some of the issues I wish to address briefly here.

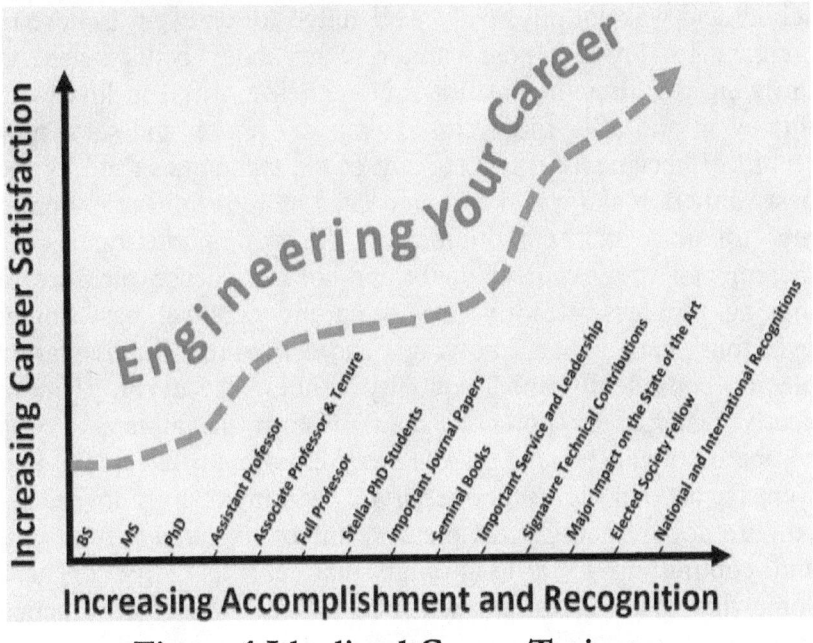

Figure 1 Idealized Career Trajectory

Of course we know that life is never as simple as depicted in Fig. 1, even with the most successful career you could imagine. There will be many local time variations in career satisfaction associated with the ups and downs of real life. Think of a great academic somewhat analogous to a great athlete. No basketball player hits every basket, no quarterback leads the team to a win every game, no runner wins every race, no tennis player holds serve every game or wins every set. Great performers are measured by "how well they play the game" over many iterations of the game itself. Furthermore, champions in virtually any sport must display a tenacity and strength of character that allows them to "shake off" disappointing losses and get in mental and physical shape for the next round of the competition. Many can go home after a disappointing loss, taking satisfaction on the aspects of their performance that were good, while drawing lessons in areas where their performance was wanting. Ideally, the "losers" constructively and realistically evaluate "why they lost" for positive motivation to significantly improve in the future.

As a specific example, I was a track and field athlete in high school and was not physically well suited for an event I chose to compete in (I was a pole vaulter with a body better suited to throwing the discus or javelin). Nonetheless, I fell in love with this event and after four years, I won my region and set a new record. I became a serious student of the techniques used by the best vaulters in the world and paid even more serious attention to my training and conditioning. Most importantly, I set incremental improvement goals and took immense pleasure at making progress, steadily improving my personal best height over four years. Then, knowing I had earned it, I finally started placing consistently and frequently winning the event. I more nearly reached my potential than most of the athletes I was competing against – and I took away important life lessons that went far beyond high school sports. Learning to truly love what you are doing, taking great pleasure in *earning steady progress*, and committing to a long-range plan are the keys. These somewhat trite suggestions will become real when you practice them in earnest. They will lead you to be the best you can be and that is after all, all you can hope for.

Intellectual horsepower, analogous to athletic ability, has a substantial genetic component. Thus some gifted individuals have an apparent "unfair advantage" when engaging in the competition implicit in modern academia. Well, deal with it. As in my high school track and field experience indicates, more nearly realizing your potential will frequently mean that you are excelling far beyond the level of your peers (if they are less dedicated to being a "student of the profession"). My experience indicates that we academics, while thinking we are working hard, are typically operating at a level far below what we are capable of, so finding room for improvement should not be too hard!

Of course the sports metaphor is imperfect, and I have likely already overworked it. However, I am confident you will agree that the set of people who hold a PhD in the sciences and engineering typically enjoy an elite level of intellectual horsepower. I am also supremely confident you have already encountered some off-the-charts bright individuals who were truly lousy professors! My experience indicates that passion and commitment to a set of important values are far more important, with regard to success as a professor, than the IQ variations amongst these highly intelligent individuals holding the requisite "union card" (a PhD).

Furthermore, there are best practices and good habits associated with being a great professor that you can acquire if you are a "student of the game," and doing the right thing can become second-nature. Regardless of your level of existing competence as a professor, I believe that you can always find room to refine your approach that will improve your performance in teaching, research and service functions. Step one: Get your mind right – that is where this book can be useful.

Engineering Your Academic Career

Chapter Two

The Precious Fluid of Life

Becoming a successful professor requires that you commit your time, your mind, and especially, your passion, to pursuit of academic excellence. This being said, it is important that you budget adequate time for the three major time blocks of your life: Work, sleep, and everything else. I recommend a nominal (1/3, 1/3, 1/3) uniform distribution, so working much more than 55 to 60 hours/week[8,9] should be considered "overtime." Overtime happens, of course when pressing deadlines arise[10]. "Everything else" includes quality time with your family, friends, recreation, and importantly, physical and mental activities of a non-technical nature to re-charge your energy. Having a well-balanced life is vitally important to long-term happiness, mental and physical health and is definitely fully consistent with being a great

---

[8] A great friend of mine, the late Aaron Cohen, spent a highly productive career with NASA, where he wore many hats, including being the head of the Space Shuttle program and Director of Johnson Space Flight Center. He was universally respected and won many national/international accolades, including being elected to NAE. He retired early from NASA and came to his alma mater, Texas A&M, to serve for a decade as a professor. In spite of his well-earned reputation as a stellar, hard-working engineer and manager, early on he was in shock at the intensity and long hours that were typical of academic life. He remarked that he just had no idea how much high quality effort faculty invest. He said he had never worked harder, been more creative, or had more fun throughout his storied career than when he taught design at Texas A&M. He also remarked that upon becoming a professor, he went from being Director at JSC where he could express a wish and have 6000 people trying to make his wish come true, to being a professor where his leverage over colleagues and the organization were, ahem, "somewhat less!"

[9] Having worked several years each in industry and government, and without denigrating either, I can tell you with authority that the hours per week spent in engineering academia exceed the other two sectors by a substantial margin, in contrast to popular (non-academic) public opinion. I know virtually no one who works 8 to 5 in academia while most colleagues did so while I worked in industry and government. Obviously many in industry and government are highly dedicated and work long hours as well, but, *it is not an implicit requirement*, as it is in engineering academia.

[10] "A deadline is an impossible due date, cleverly disguised as a great opportunity!" I believe that this cute phrase represents the unspoken management philosophy of most of the engineering industrial complex!

professor. Obviously, "loving what you do" lies at the heart of success and happiness, and should make achieving the level of effort needed feel comfortable. Being well compensated for your hobby is, well, a wonderful thing! Obviously, in reality it is not your hobby, and while fun overall, it is demanding. One reason many choose academia over industry or government is because one can stay highly technical throughout your career without being forced financially "abandon what you were trained to do" to pursue management of people and money.

Perhaps the central mental challenge of being a professor arises because the "job description" is open-ended, and there will always seem to be much more to do than you can get done. Learning to "be your own boss" when your job is open-ended is a vital step toward leading a balanced and productive life. Some qualitative considerations can help set good priorities with regard to budgeting your effort. In most academic positions, you will begin your career teaching one or two classes, and as I discuss later, you should find ways to effectively and efficiently teach and mentor students so that you have in excess of 1/3 time effort, every semester, to apply to your research and scholarship activities. Hopefully, you will be able to devote more than ½ time effort to research and scholarship early in your career. Especially during the first five years, your research and scholarship should be used as a "capacitor" to absorb available time and energy after you have taken care of other "must be done" obligations. From the onset, make a commitment to involve your students as research partners, so most of your research effort will involve mentoring of your students.

You should be conservative obligating your time, especially large blocks of effort. You can do this without coming off as an ogre if you combine your time management interaction with sound judgment and courtly good manners. An "open door" policy may sound great, when saying it to your class of 100 freshmen, but it can prove a disastrously ill-considered approach, especially during your early years. You should definitely have (generous) office hours, and students coming at other hours, without being rude to them, should be helped to understand that you will make additional time only for urgent visits.

Even though you may enter the academic world as a junior faculty member, you do not have to always defer to your department head and senior faculty by constantly saying "yes" to every request or offered assignment. Judicious use of the "yes" and "no" words are critical to your success and quality of life. You can find the balance between being a good citizen and saying "I can't get to that today," or "This week is a bear" when you are over-extended. You should likewise police carefully how frequently you say "yes" to requests to review papers, proposals, chair conference sessions, and to perform similar professional service functions. You can avoid excessive use of the "yes" word by simply standing up for yourself in a non-combative way. However, guard against being perceived as a constant "whiner," who never contributes a fair share to important mission common efforts. Common sense should lead to a balanced approach which correctly recognizes that service is frequently an opportunity and there is implicit obligation for you to perform service at some appropriate level. Service to your department may provide a way for you to display your citizenship and good judgment to the faculty at large when many of them are weakly correlated to your specialized technical interests. Performing departmental service will help give you a sense of ownership of the department and make it feel more like your professional home. However, if you never say no, you will almost certainly end up with many more obligations than you can possibly deal with. As a rule of thumb, keep your level of service functions to less than 10% of your total effort until you achieve tenure.

Likewise, don't be afraid to say no to yourself. Do not commit to pursue a large effort in a potential research project unless you are confident it is a good fit, and you are not already over-extended. Once you commit (and others are counting on you to honor your commitment), however, be stubborn and make sure you follow through tenaciously to achieve significant results from your effort. Earn a reputation for high reliability: Your word is your bond, when you make a promise or obligation, then move heaven and earth to keep your promise.

Given that you have the requisite education and desire, the main factor that will dictate your success is how effectively you invest your time. I suggest that you review your calendar at least once/week and block out a single digit number of the most important "must do" priorities and one or more tangible goals for your research progress over the coming week. This nominal agenda will be broken frequently due to unanticipated events, but keeping your plan in mind will provide the "wallpaper" for the room in which you work. Obviously you must meet your classes on the prescribed schedule; lecture-time comes whether you are well-prepared or not. I learned early on that my lecture preparation time will expand to fill all available time, plus 15%, therefore I deliberately arranged my schedule to "put myself in a generous, but not overly generous box" with regard to lecture preparation. I nominally prepare all my lectures with the formula of 2 hours of preparation for every hour of lecture time. Less time will be required for a course you have taught a few times, more time may occasionally be needed for a new course near the boundary of your sphere of competence. For the first 20 years of my career, I always tried to schedule my classes at 8 or 9 am, so that I could not avoid preparing in the evening. Upon presenting my lecture(s), I put the class out of my mind and the rest of my day was relatively clear for doing other things. Alternatively, I have found that late afternoon classes frequently gobbled all available time prior to class and that the looming lecture stayed in the back of my mind even when I was trying to do something else. So my experience indicates that afternoon lectures reduce my overall productivity. While early morning lectures work best for me, this is not a prescription for everyone.

With regard to graduate student research mentoring, it is obviously of vital importance that you align most of the research advice you give to your graduate students with the goals or background needed for each student's thesis/dissertation. Most often and preferably the student's work, even those on fellowships, will fit in well with the sponsored research projects you have in hand, or proposals you are developing, or in some cases, new areas you are trying to penetrate for future research sponsorship. You need to learn to delegate appropriate aspects of the research you are performing to involve your students. Challenge your students, but don't overwhelm them. Discuss progress frequently with student and colleague research collaborators and listen carefully to their opinions on how to redirect effort when progress is slower than anticipated. Also, talking to colleagues with even weakly overlapping research interests is often effective in breaking log jams – success is not always about the amount of time devoted. However, do not hesitate to "take yourself to task" when you are falling badly short of achieving an important goal; try constantly to work smarter as well as to re-allocate time devoted as necessary. Pick out research issues you will work on each week, correlated to the goals mapped out for students – this will advance your personal research and improve your advice to students. I recommend keeping a rough notes journal so that you can have a feedback loop around your planning process. When you hit a blind alley, record the logic for abandoning this path, and discuss the decision with colleagues before giving up – be tenacious.

I recommend you try to find a few interesting allied areas (to your main research competence areas) to work on when you grow tired wrestling with the main problem you are researching. You should spend time studying such hobby areas *not* in your present "comfort zone" spheres of competence. Actively read curiosity-driven research literature that deal with these hobby problems, even if it is just a few hours per month. I do this in a somewhat random way (however, since it is driven by my interest, I am sure it is not truly random). From these "bedtime reading" excursions, I am always on the lookout for something new that might possibly fit into my current or future "mainstream" research. I have found that maintaining some level of "technical bedtime story reading" is extremely important I work it in even when I am having a hectic week.

In mid-career, my technical hobby-reading led me to become interested in laser electro-optical sensing technology, even though I had little academic preparation or prior research that equipped me to pursue this area. I did have a need to make precise non-contact measurements in my work, however, and I realized that "the speed of light is truly a wonderful thing," especially as I came to appreciate some of the recent advances enabling precise modulation of laser power, photo-detection, time keeping, and signal processing. Collectively, these advances in turn have enabled a plethora of new devices, limited only by our imagination. Although I had little background in photonics, I saw several needs where new electro-optical sensors appeared to be feasible that offered opportunities for truly significant advances. I soon "learned enough to be dangerous" and became obsessed with the rapid progress I could see being made in the theoretical and applied research literature as well as the commercial devices that were emerging. Through such explorations, I soon developed a passable level of competence and perspective on this new (for me) area. At some point, I began doing conceptual designs that ultimately resulted in novel optical sensors and algorithms for processing the resulting measurements. My experimental research now routinely makes use of advanced optical sensing, some of which I have invented. These technologies give a measure of uniqueness to my laboratory and are a source of great satisfaction. Even though I do not claim to be a world authority, I find that tracking the rapidly evolving technology and making occasional significant contributions in this field are highly stimulating to me and my students. My experience indicates that serendipity is frequently "earned good luck" – and my occasional eureka moments make it clear to me how important interest-driven "pleasure reading" can be. In fact, I believe the majority of my most significant research contributions are traceable to "pleasure reading." Even though you may engage in such studies just for the fun of it, I believe you too will find these efforts will pay a large dividend in your research. So cultivate technical hobbies, for fun and profit! Finally, I am fairly confident these technical hobbies will help keep you technically young and maintain your ability to move smoothly in new research directions as technology evolves over the course of your career.

Chapter Three

What Kind of Research Do You Want to Do?

You should already have some qualitative career research goals in mind, especially for the first few years. Do you wish to develop a specialization in some aspect of engineering science – emphasizing theoretical formulations? Or, do you wish to specialize in experimental methods, computational methods, or design? Do you wish to develop a set of competencies and strive for a high degree of technical breadth? It is important to balance breadth versus specialization, and recognize that research breadth is most often built on a sound education in fundamentals and reinforced over time by integrating across a set of specialized research experiences. The theory *versus* experiment mindset should, I believe, be replaced by theory *and* experiment, in order to achieve the balance that enables efficient transition from theory to experiment to validation, and ultimately, to realizations of your ideas as advanced technology.

To help focus this discussion, Figure 2 gives a depiction of the career contributions of four notional professors. Two of the distributions represent the contributions of late superstar Professors Willard Gibbs and Theodore von Kármán. Both made landmark contributions of eternal significance that dwarf most professors' lifetime contributions. However, observe that their distribution of contributions and competencies were very different. Professors X and Y represent career contributions of two strong modern-day professors with contrasting sets of contributions. Professor X's main strengths are in theoretical and analytical methods with some contributions to experiment and design. Professor Y is primarily an experimentalist who has competence in design, and who has also made some theoretical contributions.

The curves in Figure 2 are for qualitative and heuristic discussion purposes. We know that an individual's technical contributions, to the extent they might be represented meaningfully as a continuous one-dimensional distribution function, would certainly not be a stationary process in time. We should expect the areas and nature of research to evolve, perhaps varying substantially over one's career.

There are always risks of making strong inferences from historical examples as a guide to the future. In this text, the risk is not that historical analogies are irrelevant. Rather, the risk is that the two individuals discussed below, Willard Gibbs and Theodore von Kármán, stand so tall that they cast too long a shadow. They are discussed not to intimidate, but rather to illuminate. We should not set these giants as the standard by which we measure ourselves (because the comparison will be pretty embarrassing for most of us!), however, we can use Gibbs and von Kármán as motivating examples of how truly outstanding academic careers can be pursued with sharply contrasting emphasis on theory, technical breadth, experiment and design. I hasten to add, the interested reader should pursue further study to do justice to these two remarkable individuals.

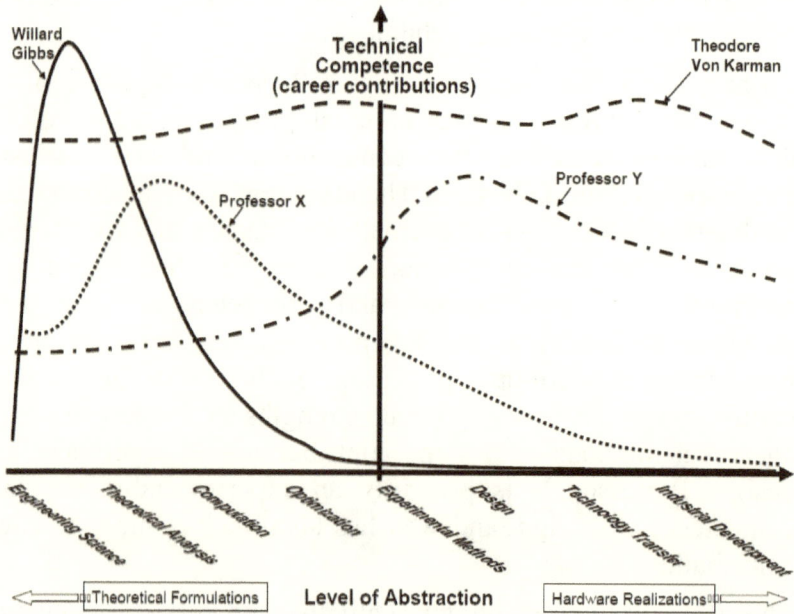

Figure 2   Career Contributions and Technical Competence Distributions for Four Notional Professors

Willard Gibbs (1839-1903) was awarded the first U.S. PhD in engineering (Yale, 1863), moreover, he made monumental theoretical contributions in applied mathematics, physical chemistry, and mechanics. His work remains alive today and is important in virtually all branches of engineering and applied science. He was the prototypical engineering scientist. While his work was essentially done without recourse to laboratory research, there is no doubt that his work was informed by the experimental research done by his peers.

Gibbs' innovations are summarized in reference [2]. Among the most influential of his contributions are

- Vector algebra and calculus
- A rigorous mathematical basis for thermodynamics
- A rigorous electromagnetic theory of light
- Insights in approximation theory and in particular, Fourier series convergence
- Novel methods for orbit determination

Among the important results named for Gibbs insightful contributions are the following:

Gibbs free energy, Gibbs phase rule, Gibbs entropy, Gibbs inequality, Gibbs phenomena, Gibbs paradox, Gibbs-Helmholtz equation, Gibbs algorithm, Gibbs distribution, Gibbs state, Gibbs sampling, Gibbs preliminary orbit method, and Gibbs-Marangoni effect.

Gibbs' name belongs to the same set with European giants Brahe, Kepler, Galileo, Newton, Euler, Lagrange, Laplace, Poisson, Legendre, Jacobi, Gauss, Rayleigh, Cayley, and Hamilton. Over three centuries ending around 1900, these European giants developed the fundamental insights and basic results upon which virtually all of engineering and applied science presently stands. Our debt to these immortal scholars cannot be overestimated. While European scientific elites had long admired American technological creativity through the monumental inventions of Bell, Carver, Eastman, Edison, Fulton, McCormick, Morse, and

Whitney; Willard Gibbs was the first American scientist/ mathematician/engineer to make contributions of such fundamental gravity that his name fits comfortably in the same paragraph with Newton and Gauss.

It is impossible to overstate the importance of Gibbs' contributions to modern engineering and applied science. His vector algebra and calculus dominate the language of mechanics in physics and engineering. Gibbsian thermodynamics remains fundamental for the thermal sciences and engineering. His contributions to the physics of light remain highly significant milestones.

While Gibbs worked in the left half plane of Figure 2, he drew insight and motivation from challenges posed by problems he saw across the spectrum from basic science and mathematics to applied engineering. His work led to results of such significance that they still provide pervasive foundational relevance across the spectrum from basic theory to applications. Indeed, Gibbs' contributions, over a century after his death, continue to give credence to the axiom that there is nothing more practical than good theory.

Willard Gibbs was a tough act to follow, but the 20th century brought us many new heroes and a wide array of rapid developments in science and engineering. Let us consider a second icon of excellence to draw further insights. Theodore von Kármán (1881-1963) was an engineering professor whose breadth and depth of contributions are difficult to capture briefly. They can be pursued further in the references [3 – 9]. He was among a small group of 20th century contributors whose work mapped the classical mechanics of the 19th century into the foundation for modern engineering. No one can study structural stability, aerodynamics or propulsion without encountering von Kármán's important innovations and pervasive influence. His impact was amplified tremendously because he combined basic theoretical and experimental research in structures, aerodynamics and rocket propulsion with amazing technical breadth and a full range of applied engineering knowledge. Remarkably, he integrated these in several ambitious design-build-fly projects executed at California Institute of Technology. The philosophy implicit in his career is worthy of further evaluation.

From early in his career, von Kármán showed a burning ambition to achieve theoretical breadth and depth. He sought to acquire a broad set of engineering knowledge that would allow him to quickly move from concepts to experimental studies and hardware realizations. After winning the Hungarian Eötvös national prize in mathematics as a teenager, he entered the University in Budapest to study mechanical engineering.

Von Kármán excelled at every level and his doctoral dissertation (1908 under Prandtl at Göttingen), where he made landmark contributions in structural mechanics (a theory for stability and elastic/plastic buckling). During his graduate studies, however, he pursued mechanics education and research broadly, completing enough original work that he could have written a second dissertation in fluid mechanics. Indeed Prandtl considered von Kármán his heir apparent in theoretical fluid mechanics.

Although von Kármán's laboratory skills were initially his weakest suit, he placed tremendous value on iteration between theory and experiment and found ways to make it happen. Even when his experimental skills were not well-developed, he could not stand to spend his life making a sequence of hypotheses, and building elaborate theoretical structures on the hypotheses. He would find himself asking frequently how he could test the theory experimentally to learn which of the hypotheses agreed with physical reality. So he began to form small research teams, collaborating with one or more experimentalists on jointly designed experiments. He was an early embodiment of the elusive present-day talisman, an interdisciplinary researcher. Over the early part of of his career, he vastly expanded his own experimental research competence and emerged as a top experimentalist, while at the same time expanding his portfolio of basic engineering science results. His focus moved eventually toward combined theoretical and experimental studies in full up systems such as airplane aerodynamics and rocket propulsion, then rocket flight mechanics, and finally, he turned to invention and design. Ultimately, he engaged in large applied research projects to implement his theories in flight tests, and he became the premier international aeronautics and astronautics researcher during the 1940s and 1950s.

As documented in his Collected Works [3], von Kármán's contributions in theoretical fluid mechanics were extensive and established him as a world-class theoretical authority in aerodynamics. His work documented fundamental innovations and insights in unsteady wakes due to flow past a cylinder; stability of laminar flow, theory of turbulence, airfoil theory in steady and unsteady flow, boundary layer theory, and transonic and supersonic aerodynamics. In the literature, his name is associated with the following theoretical aerodynamics contributions [3]:

> von Kármán vortex street (flow past cylinder), von Kármán integral equation (boundary layers), von Kármán-Pohlhausen parameter (asymptotic expansion for boundary layers), von Kármán-Tsien compressibility correction, von Kármán Ogive (supersonic aerodynamics), von Kármán-Treffz transformation (airfoil theory), von Kármán-Nikuradse correlation (viscous flow), Prandtl-von Kármán law (velocity in open channel flow), Chaplygin-Kármán-Tsien approximation (potential flow), Falkowich-Kármán equation (transonic flow)

He also made a remarkably wide-ranging set of other engineering mechanics contributions, including a lattice model for crystallography and a variety of contributions in solid mechanics. Prandtl at Göttingen and von Kármán at California Institute of Technology were the two dominant ancestors of modern academic research programs in mechanical and aerospace engineering.

While he was very gifted in both basic and applied research, von Kármán's ambition to see real-world engineering impacts flow from his analysis led him to pursue ambitious design-build-fly projects: from gliders to helicopters to jet airplanes to rockets. His innovations truly changed the world; von Kármán does not have a close competitor as regards which academic figure was most responsible for the jet and rocket age. It is truly amazing that he also found time to pursue research while simultaneously performing equally impactful leadership functions.

He was head of the Aeronautical Institute at the University of Aachen (1912-1930) where he designed and supervised construction of their first wind tunnel, and he directed the Cal Tech Guggenheim's Aeronautical Labs at Pasadena (1930-1957), designing and bringing to reality Cal Tech's first wind tunnel and propulsion laboratories. Von Kármán was also a prolific and highly effective mentor. He formed a strong bond with his former PhD students, who were known as *von Kármán's Circle* (e.g., Malina, Summerfield, et al). He earned a reputation as a zany, humorous professor who was simultaneously a genius in physics and mathematics, as well as an effective project engineer who could quickly take ideas from his lecture blackboard to successful flight projects. His reputation as a stellar classroom teacher and his ability to develop excellent graduate students greatly amplified his research effectiveness and his legacy. He and his former student Frank Malina led the "Women's Air Corps" (WAC) Corporal project. This was the pioneering US Army project around which the Jet Propulsion Lab was formed with von Kármán as the first Director. Malina was the second Director.

The WAC Corporal label for their rocket was rather curious, apparently it referred to the fact that the WAC was considered, qualitatively, a "little sister" rocket whose "big brother" was the larger Corporal rocket under development and test by the U.S. Army at that time. The Women's Air Corps was a novelty during the mid-1940s, and in researching this rocket, I have not uncovered with certainty the origin of this name, but I believe it to be an invention of either von Kármán or Malina. Von Kármán and Malina launched the WAC Corporal rocket and it flew to the world altitude record of over 227,000 ft in 1945, shattering the records of von Braun's V2 rocket. Von Kármán and Malina also founded, in 1945, the first jet and rocket engineering company (Aerojet General) that was destined to play a leading role, now approaching seven decades, in developing, implementing and commercializing jet and rocket technology.

During the final 20 years of his career, von Kármán also proved to be a remarkable national and international organizer and leader. He chaired the first Scientific Advisory Board to the Chiefs of Staff (1946, whose first report led to the creation of the U.S. Air Force and the dismantling of the Army Air Corps). This scientific advisory organization was re-named the Air Force Scientific Advisory Board and continues to give transformative technological advice to the Air Force to this day. He was the founding Chair of the Advisory Group for Aeronautical Research and Development (AGARD, 1951), which has since then been NATO's aeronautical research arm. He led the formation and was the first President of the International Council of the Aeronautical Sciences, ICAS (1956). Finally, he also was the founding President of the International Academy of Astronautics, IAA (1960). The IAA and ICAS remain influential international organizations. Von Kármán's awe-inspiring breadth and depth of research and teaching contributions, time-shared with phenomenally effective leadership was unparalleled in his lifetime and has not been approached since. In summary, he was the dominant figure that led the transformation from propellers to jet propulsion, and he brought us to the dawn of the space age.

Von Kármán was famous for his well-developed and sometimes biting sense of humor. While he retained appreciation of theoretical analysis throughout his career, he liked to poke a little good-natured fun at "pure theoreticians" whom he felt frequently made assumptions for the sake of mathematical convenience or elegance, as opposed to best match physical reality. He once humorously defined a theoretical aerodynamicist (and he was one!) as "An individual willing to assume anything but responsibility!" In another classical example of Kármánesque wordsmithing, he distinguished an engineer from a scientist with the following quote (my paraphrase): "A scientist studies the world as it is, an engineer creates new worlds."

Some of von Kármán's humor is not politically correct these days. While I will largely not "go there" on his humor, perhaps one self-deprecating ethnic joke he told on himself is in order. Von Kármán was Hungarian, and while he was generally well-liked and gregarious, he earned a reputation for occasionally being "just a little" overly-aggressive. He once defined a Hungarian as "Someone who can enter a revolving door behind you and exit in front of you!" We must always evaluate humor and political correctness in the light of the times in which people live. Von Kármán was appreciated by most who worked with him as a wonderful human being, generous in spirit, but with a pattern of mischievous teasing in interactions with his students and colleagues which occasionally went "out of bounds."

Implicit in von Kármán's witty sound bites, and especially in the examples set by how he quickly moved his own research to maturity, one can infer his technical philosophy. I am virtually certain that von Kármán would recommend that analytical/theoretical engineers spend at least some fraction of their effort coming to grips with any gaps that may exist between their formulations and what is necessary to make their ideas actually impact the world as advanced engineering realizations. While most of us can only dream of a career analogous to von Kármán or Gibbs, mapped into a modern setting, we certainly can take inspiration from their examples.

We should feel especially humble when we observe that von Kármán's and Gibbs' contributions were done during an era sans computers, sans word processing, sans image processing, sans user-friendly systems for data acquisition, sans sixteen bit a/d conversion, sans computer-automated experiments, and sans the internet.

So what does a young professor take away from this discussion and Figure 2, other than being impressed, inspired, and possibly intimidated by Gibbs and von Kármán? The most important message is good news: There exist many ways one can pursue a successful academic career. Also of significance, it is not necessary (or desirable) to "label yourself for life" with a sharp specialization, but rather anticipate that your career will likely evolve in exciting ways that mimic on a more modest scale the distributions of von Kármán or Gibbs. Some specializations and focus are necessary for progress during your early years. I am sure it has not escaped your attention that it takes a few years to "own" a sub-discipline. Breadth and depth as graphed in Figure 2 cannot happen in a day or a month. Rather depth and breadth flow from extended strategic investments of your effort over decades.

As your career unfolds, I recommend you stay on the alert for opportunities to move your more promising ideas and results toward maturation and near term utilization. Challenge yourself to reach outside your main existing strengths and do what is necessary to truly advance your work and demonstrate implications for advancing the state of engineering practice. Simply "throwing abstract papers over the ivory tower wall" in an intermediate state of maturity, and implicitly praying someone broader, or someone with more initiative, or that some appropriately structured organization will adopt your ideas and complete what you started, ... well, sometimes such prayers are answered and sometimes not! On the other hand, if you prove to be a Gibbs class theoretician, you may not need to be overly concerned with personally engaging in experiments or collaborative efforts that make things happen in the right half plane of Figure 2. The power of your ideas and the clarity of your presentations may prove sufficient to cause constructive right-half plane evolutions to happen "for you". If after a few years, however, your results are not gaining the traction needed to make a constructive difference in mapping your conceptual and theoretical research into significant impact on advanced engineering practice, you might draw lessons from the von Kármán example and take the necessary personal initiatives that move your work toward maturation and greater near-term impacts.

It is evident that von Kármán was a sufficiently strong analyst that he could have enjoyed a high degree of success in the left half plane of Figure 2 by "working less hard" and restricting his effort to researching the mathematics and mechanics of fluids and solids. On the other hand, I believe he would have had a lot less fun, and looking carefully at his monumental accomplishments, we can see the incredibly larger impact he made by adopting a much broader stance. He invested tremendous energy to develop analytical, computational, experimental, design and management skills, and the ability to fuse them all together in impressive projects. There is no doubt he chose to pursue both breadth and depth to maximize the influence of his ideas and because he loved seeing closure between theory and physical realizations. His amazing ability to move his ideas from theory to computation to design and in many cases, to successful flight demonstrations, maximized his central role in the 1940-1960 technological revolution in air and space flight. One can only imagine the satisfaction that he must have felt when his WAC Corporal rocket flew to a world record altitude. Occasionally, on a much more modest scale, I have personally observed that effort focused on advancing the maturity, insights, relevance and utility of basic research leads to important new conceptual advances that would not have happened otherwise. In other words, basic research can often be significantly accelerated if time is taken to mature some of the central ideas and test their practical implications.

It is evident that von Kármán's impact and influence were dramatically advanced by his technical breadth and ability to integrate across a wide range of analytical, computational, experimental, design, and implementation issues. The key to unlock creativity and maximize impact is very individualized, however, and the freedom to find one's own way should be retained. We should hope for numerous future disciples of both Gibbs and von Kármán.

Related to the above discussion, I frequently participate in the nomination and selection process for the National Academy of Engineering (NAE). A challenging aspect of documenting an outstanding academic researcher's candidacy is to show how his or her work has made "real-world" impact on the advanced state of practice in the field. Impact, in a critical evaluation, should assess not only impact on the research literature, but also how the

31

candidate's work has advanced the state of engineering practice and/or led to advanced technology that has been used in important engineering realizations. This broad definition of impact seems to be a reasonable standard for the highest national honor for accomplished academics (who have spent about three decades in research, scholarship, and teaching, since the average age of induction into the academy is around age sixty). This perspective perhaps favors imitating the von Kármán career model as the most accomplished example. However, there is no doubt that Gibbs' work made much more than enough impact during his life to far exceed this standard, and I am personally aware of NAE members who very successfully pursued the Gibbsian approach. This perspective leads to the summary statement: The highest degree of professional success for an engineering professor, is achieved when one has authored fundamental advances and it becomes evident to peer colleagues that the contributions of the individual and his or her team of collaborators have been realized in important advanced technology or impacted the practice of engineering, during the person's professional lifetime.

I hasten to add, some colleagues differ with this point of view, saying this perspective forces research to be too applied. They might further argue that publishing a sufficient volume and quality of analytical/engineering science results in the archival literature is both necessary and sufficient to achieve the highest degree of success (i.e., no expectation that the scholarship be realized as advancements in engineering practice). Certainly both Gibbs and von Kármán conducted basic research, disseminated their results in the literature, *and* their work made most exceptional impact on the state of engineering practice during their lifetime.

It could be argued that von Kármán adopted a much broader stance from the onset and was more aggressive in maturing and applying his own research results. On the other hand, considering the success of Gibbs' engineering science and mathematics in very dramatically changing engineering practice, he obviously gets very high marks on the engineering practice relevance of his research. The contrasting Gibbsian and von Kármán approaches to engineering are both "highly respectable" today among engineering academics. When disagreements arise among senior faculty regarding the merit of young faculty research accomplishments, it is quite frequently about what relative value the Gibbsian or von Kármán models.

In a utopian academic universe, professors with a particular distribution of technical competence would maintain a broad appreciation of the full range of needed competencies. The "technical value system" we bring to the table should not necessarily correlate with our own areas of competence or demands we make on ourselves. Rather we should bring a perspective that reflects the broader set of competencies needed in our ever-changing fields of engineering. The faculty technical value system will not be uniform across any department. However, from the perspective of the younger faculty, the views held by their senior colleagues are clearly important because they affect decisions on recruiting, promoting and hiring.

I would hope that most would agree that both Professors X and Y depicted in Fig. 2 are to be highly valued and their careers should not be damped significantly by technical prejudices of their senior colleagues. Specifically, experimentalists and design-oriented professors should retain appreciation for the value of more analytical/abstract researchers, and vice versa. For example, while Gibb's competence distribution did not include the experimental expertise, design, and system realization strengths of von Kármán, it is virtually certain that he would have recognized a young von Kármán's promise and would have anticipated and appreciated the enormous impact that von Kármán could make. I am sure that there are lessons here for us lesser mortals.

Young professors can be comforted by the truth that senior faculty of differing "technical DNA" engage in spirited discussions of these issues during tenure, promotion, and hiring decisions. In my experience, these deliberations filter out most potential miscarriages of justice to reach good decisions. In addition to affecting the careers of younger colleagues, senior faculty recommendations on tenure, promotion, and hiring represent investments in the future of their department. The senior faculty generally approach these decisions with the gravity they deserve. Judging from most nationally prominent engineering departments I know well, this process does indeed lead to career advancement for most strong professors whose technical strengths correlate qualitatively with both X and Y of Fig. 2.

Assistant Professors have important decisions to make early in their careers. I believe the direction of early research effort is amongst the most important. Your decisions on research avenues to pursue should be driven by several considerations:

- Your current technical competencies
- Your most passionate research interests
- Your assessment of technical and economic trends, needs, and funding outlooks, including immediate opportunities (collaborators, facilities, research support)
- The technical prejudices of your senior colleagues.

While you, as a young or mid-career faculty member, should hopefully not need to stay up nights worrying about the last bullet, you should also not ignore it altogether.

Well, back to Earth – where was I? Oh yes, I am giving constructive advice as regards having a successful academic career. In the next section, I provide some statistical metrics derived from an informal survey of 15 nationally ranked engineering departments. These data are provided not to establish hard constraints, but provide typical "volume of output" measures as a rough guideline.

Chapter Four

Recent Professor Promotions Statistical Data

Metrics for professors recently promoted are provided in Tables 1 and 2. These data are collected from all promotions in fifteen top 15 engineering departments over a six year period, from the following institutions: MIT, The University of Texas, Purdue, and Texas A&M University. Over half of the departments whose data was considered were top ten departments as ranked by the US News and World Reports. The data were collected from administrative colleagues at these institutions who wish to remain anonymous. Care has been taken not to violate the confidentiality of the individuals, nor can their departments be identified. The statistics are not comprehensive, only a small subset of the departments from these institutions participated. However, most disciplines were represented and the sample size is large enough to give useful insight. Nonetheless, these tables do not representative a scientific statistical analysis and should be consider as representative.

Table 1 summarizes data from promotions from Assistant to Associate Professor, with tenure. All averages have been rounded off to the nearest integer. Notice the broad generalizations: Promotion and tenure occurs when the junior faculty member has graduated an average of 6 MS and 1 PhD students, published an average of 15 journal papers and presented an average of 26 conference papers. The successful Assistant Professor's publications average over 50 total citations (according to Science Citation Index), with an average of 2 of their papers drawing 10 or more citations by their peers. Successful candidates are typically offering 1.25 courses per semester and are advising an average of 4 graduate students.

Table 1 Statistics for Promotions from Assistant to Associate Professor (with tenure) Sample Size: 61

Statistic	Journal Papers	Conference Papers	Total Journal Citations	Papers with >10 Citations	MS Grad uates	PhD Grad uates
Mean	15	26	56	2	6	1
Maximum	24	40	90	4	10	3
Minimum	7	11	15	0	2	0

Table 2 Statistics for Promotions from Associate to Full Professor Sample Size: 40

Statistic	Journal Papers	Conference Papers	Total Journal Citations	Papers with >10 Citations	MS Grad uates	PhD Grad uates
Mean	34	50	212	7	10	7
Maximum	82	130	900	19	18	11
Minimum	14	24	57	1	3	3

Table 2 provides analogous data for promotions to Full Professor. The average volume of published scholarship is substantially increased to 34 journal publications and 50 conference papers. The citation count is substantially increased with an average of over 200 citations and with 7 of the journal papers having a double digit number of peer citations. The graduate student degree completions are also significantly increased, with promoted Full Professor candidates having graduated an average of 10 MS and 7 PhD students. Full Professors average teaching about 1.5 courses per semester and advising about 5 graduate students at the time they are promoted. Notice the wide variations in all of the metrics, especially the citations. It is noted that the citation data in Table 2 are skewed because one candidate had 50% more publications and almost three times the citations of the closest peers. While the increased average numbers of journal conference papers, journal papers, and citations are consistent with the increased expectations for promotion to Full Professor.

It is noted that the journal publication and citation data vary widely by field, with long-standing disciplines where there exist a large population of active national/international peers generally resulting in large citation counts. There are certain areas of specializations where only a small single digit of journals are appropriate to publish a specialized paper, whereas, other papers may produce more universal methodology with tens of journals where the paper fits. Obviously papers of the latter variety would have greater opportunity to generate a large citation count. Many non-engineering fields, for example, chemistry, physics, and biomedical science generate exceedingly high citation counts compared to engineering, because of the size of these disciplines and the traditions for publishing many shorter multi-authored papers. For example, a routine paper in a biomedical field generates hundreds of journal citations, whereas in many engineering sub-disciplines many highly accomplished members of the National Academy of Engineering may have very few papers with more than 100 citations. Certain other engineering disciplines, for example Computer Science, do not emphasize journal papers but rather conference papers, where rigorous peer reviews are applied to the most prestigious conferences. Because of this lack of consistency with other engineering disciplines, no Computer Science promotions were included in Tables 1 and 2.

The data in Tables 1 and 2 certainly measure the volume of scholarship, and the number of archival papers and the citation data do also provide a measure of the quality and impact of the candidate's work on the literature. One important perspective is that careful evaluation of citation data must can be done with a view toward assessing "market penetration," taking into account the number of papers published in the particular candidate's areas of specialization. The data in these tables do not take these issues (size of the discipline, traditions of the discipline, and the "market penetration" of specific disciplines) into account.

The quality of teaching, research education and mentoring certainly are not captured in Tables 1 and 2 and, indeed, are difficult to measure without being well-positioned to observe the individual candidates. While the data in these tables, when properly qualified and understood in the context of the individuals' fields, do provide good measures of impact on literature, they typically do not measure well the following:

- the impact of the candidates' work on advancing the state of advanced technology actually adopted in their fields

- development of important laboratory facilities

- invention of novel devices

- development of new courses,

- innovations in teaching

- quality/quantity of service and leadership functions

All of that having been said, in my judgment, it would be unusual for a candidate to be promoted whose volume of scholarship and student production did not significantly exceed the minima for most metrics in Tables 1 (Associate Professor with Tenure) and Table 2 (Full Professor). A few exceptional areas of high productivity, however, together with well-informed and documented assessments of high quality in some areas, are frequently sufficient to offset a few low metrics.

Instances occasionally arise in which most of the average measures in Tables 1 and 2 have been significantly exceeded, yet a candidate is not promoted or tenured. Likewise, some candidates with volume numbers significantly lower than many of the metrics tabulated are promoted and tenured. Such apparent statistical miscarriages of justice beg the obvious question: "What does it take?" Put in the simplest terms possible, the professor's overall contributions must be of sufficient quantity, quality and impact to merit the promotion, in the judgment of senior colleagues in their field, and especially the Departmental Tenure and Promotion Committee. For example, it is possible to have most of the metrics well above the averages and have a poor record mentoring graduate students. An un-tenured professor with strong metrics otherwise, but with poor graduate student mentoring record may not be offered tenure in a program that strongly weights development of MS and PhD students. These judgments are brought to an initial focus at the department level. Senior colleagues well positioned to evaluate the candidate's work also have the earnest responsibility and strong incentive to make fair decisions affecting the future vitality of their department. Strongly positive recommendations at the departmental level usually carry sufficient gravity that the college and university reviews usually ratify the department's judgment.

It is very important to note that research which produces advanced devices must be evaluated in other ways that take into account of the quality and impact of the individual's output. For example, Jack S. Kilby was awarded the Nobel Prize in 2000 for his contributions to the invention of the integrated circuit (mid 1950s). He published a career total of four journal papers, and these received to date a total of 41 journal citations. All 41 of these citations are of his retrospective article reviewing how his research effort led to the first integrated circuit, written over twenty years after he completed it! Yet the pervasive importance and impact of his work on the evolution of the computer industry are widely recognized as being of impossible-to-over-estimate significance. Of course he performed this research over several years in an industrial laboratory setting at Texas Instruments where his research progress was evaluated first-hand in his laboratory (today, we would recognize this laboratory as the keystone pioneering computer engineering laboratory), using experimental metrics that have no correlation to counting journal publications and citations. Kilby's work was documented in patents and company reports, of course, but these publications "don't count" as archival journal papers.

The obvious truth that Kilby's research quality and its phenomenal impact are not captured by journal metrics should not escape our attention, even given the obvious fact that Kilby was not a professor. Some would argue that he was not a scholar according to an engineering science definition. While there is some basis for this observation, Kilby was clearly an important engineering researcher whose contributions impacted in a pervasive way nearly all aspects of modern engineering practice and, indeed, changed the world. Most of us can only dream of making such an impact. In other words, academia is not the only game in town as regards research and innovation. Estimating the quality and impact of an individual's contributions based on publication metrics has some evident limitations.

Journal paper and citation metrics are especially weak in measuring quality when research results in technological advances realized as inventions or devices. The research process leading to inventions frequently does not generate a conventional paper trail of journal articles so the work can be "graded" using our narrowly tailored metrics. The citation metrics are obviously most constrained when the invention is considered intellectual property of the organization where the work was performed. Given the present push by most major universities to increase the volume and quality of inventions, intellectual property, and technology transfer through licensing and high technology spinoffs, this observation is especially relevant.

We need to look carefully at the importance of research contributions from multiple perspectives. We surely would not wish to cultivate an environment with strong disincentives for the next Jack Kilby, just because he or she happens to be making inventions and doing highly innovative research and development in a university environment, would we? In the reality of the present university environment, young Kilby would be well-advised to publish a "passable" number of refereed journal papers, consistent with Tables 1 and 2. It is hoped that his senior colleagues, who were well-positioned to see his work in the laboratory, could develop an appreciation for the importance of his inventions of novel devices and give these full consideration in evaluating his performance.

Certainly young professors should take note of the above data, and the observations regarding the challenges we face in interpreting these metrics. I recommend that you avoid approaching your task as "just feeding the meritocracy some big numbers, because it is just a numbers game." It is far more vital to understand that optimizing the actual quality of your teaching, research, scholarship, and service is the key to success. This is especially true if you take the long view on having a successful career. In hindsight, I can see several of my early papers that I would like to "un-publish," I am glad that there are only a few of them! Publishing is a lot easier and much more fun when you have something really new and significant to report. While quality scholarship may be difficult to plot on a graph, most of us can recognize it when we see it. So you should have a little faith that truly significant research will be recognized and valued in some way that goes beyond a bean counting mentality. This should especially be true amongst your departmental colleagues who are in a position to judge the degree of innovation in your work.

In a well-functioning department, a few exceptional, ground-breaking papers will trump a large collection of pedestrian, incremental publications. While serious effort is usually made to evaluate the quality and impact of scholarship during tenure and promotion deliberations, it is important to note that these judgments always carry a degree of subjectivity, especially above the department level. It is therefore vital that you to strive to generate a reasonable, at least acceptable, volume of journal publications, and always optimize quality. In mentoring young faculty, I encourage them to push hard to do something important in every paper submitted for publication and hold off publishing until the work meets their standards for excellence. Given a reasonable number of publications, your reputation for excellent scholarship will be based on the quality of your work, not the volume.

In spite of the fact that I risk discouraging, to some degree, the next Jack Kilby, I would recommend against untenured professors placing a high priority on generating proprietary technologies and patented inventions. Establishing basic research credibility and visibility as an academic should be a top early priority. While I am certainly no Jack Kilby, I did my first patented invention at age fifty – this is obviously unusually late if one wishes to have a distinguished career as an inventor. On the other hand, I have really enjoyed this change of pace, and have since come up with a significant number of good inventions. In my case, having strong accomplishments and experience in basic and applied research proved advantageous in all of my inventions to date.

My late blooming interest in inventions does make the point that you can generate patents after completing the tenure process. Upon gaining tenure, you will still be young enough to time-share many inventions if this is an area of strong interest. I would therefore advise un-tenured professors to not emphasize patents and inventions, but invest their effort in developing publishable research and submitting frequently to the main journals in their areas of interest. This advice notwithstanding, should you come across an idea for a potentially important invention, by all means submit the patent disclosure immediately. The primary outlet for documenting academic research results for young professors seeking tenure will remain, for the foreseeable future, the archival literature in the appropriate field.

Publishing fifteen or so papers in good journals carries with it the judgment of several dozen reviews by peers who have evaluated multiple examples of the young professor's work and concluded he or she has made a sufficient contribution to warrant publication. So volume of publication in rigorously reviewed journals is a significant measure of research quantity and quality, as are positive citations of your published work by peer researchers. Therefore publishing frequently in good journals is the surest path to accumulating clear evidence supporting the quantity and quality of your research contributions.

Beyond imposing some minimal publishing expectations, one should strongly push to perform research targeted on truly significant advances in the field. Your reputation in the field will typically be dominated by your peers impressions of your journal and conference papers quality. You only get one chance to make a first impression, and every paper you write will introduce you and your work to new colleagues, and you should keep this in mind. Your collection of publications will someday represent the main documentation of your technical legacy.

Writing in a logically cohesive manner with minimal "snow jobbing" is really important. Obscuring physically easy to understand ideas through pages of obtuse and overly abstract mathematics can be a real turnoff, unless the mathematics is truly necessary from a conceptual point of view. When I read a paper in an area I know well, it is trivially easy for me to see through smoke screen language with un-needed abstractions that attempt to promote a small molehill of a contribution into a Mount Everest. Nothing more quickly turns me off and, it makes it hard for me to begin reading another paper by that author with a neutral opinion. Develop a reputation for doing rigorous analysis, but do not forget that (following Gibbs example), you can and should harness rigorous, but readable, mathematics to establish widely accessible insights. If your work is significant, a clear, minimally abstract presentation will give you maximum impact.

Your claims as regards generality and impact should, perhaps, be mildly understated; this style will always earn my respect. If your papers are to have real impact, some of your readers will invest heavily trying to build on your results. So do your best to ensure the reader who attempts to repeat what you have done will find the effort worthwhile and be able to not only verify what you have published but also find it a useful foundation for further work. Again, we can and should draw a clue from Gibbs and von Kármán, either of whom could have written their papers with a level of mathematical abstraction accessible by only the most stellar and motivated theoretical and mathematically literate scientists of their day. Had they pursued this path, it is doubtful we would view them today as icons, and their names would not be "household words" in engineering, and obviously, their careers would have virtually certainly have made much less impact. More to the point, their whole lives were clearly devoted to doing great research, making monumental engineering advances, and communicating the results in a fashion that allowed near-immediate access by, and impact on, the widest audience possible. Once again, there is a lesson here for us lesser mortals; pay attention to this issue and your life will be more rewarding and a lot more fun.

An important challenge a young professor faces is striking a balance between making long-term research investments versus producing near-term results for publications. This tension may be acute for experimentally-oriented researchers requiring difficult-to-develop experimental facilities. This tension is natural, and the best advice I can give is "deal with it!" One approach to consider, think about your long-term research goals as part of a strategic plan; I believe you will usually find that long-term strategy and publishable short-term research planning are not mutually exclusive. If you are careful in putting together your long-term plans, you will likely find that you can generate significant intermediate results for publications, and if not, short-term research activities should be pursued in parallel. Since your graduate students need to make significant progress within a year or two, you need to plan your work and potential journal publications with compatible milestones. Using such thinking, research results and journal papers should flow at a reasonable rate, even while you are pursuing a multi-year plan. Young professors should vet their approach to these issues and multi-year plans with senior colleagues to help find the right balance.

Chapter Five

Grantsmanship:
Why All the Pressure Regarding Research Sponsorship?

Among the most difficult and perhaps most unfamiliar challenges faced by a newly minted Assistant Professor is the *M word*, Money for research sponsorship. In the hindsight evaluation of successful professors, I suspect most would agree that the most important issue underwriting their eventual success was actually the quality of their research, mentoring, and teaching. However learning how to succeed in proposal-writing is by no means trivial.

Analogous to publishing journal papers, research proposal-writing is easier and far more effective, and fun, after you have successfully performed some innovative research. To write a successful proposal, you must begin with a significant set of research results already in hand that targets issues lying in the intersection of your areas of interest and the set of currently fundable research areas. Your prior research results will dominate the considerations that underlie the review process. A proposal that outlines an aggressive program promising novel results in an area where the author of the proposal has little or no historical research accomplishments is usually dead on arrival. So it has been said: "Riding the sponsored research train is a lot easier after you successfully board it!"

Most beginning professors can "better get their minds around" the teaching and scholarship roles because they can readily extrapolate what they need to do from their experience in graduate school or as a post-doctoral researcher. Acquiring research support is obviously very important, however, because it enables the stepping stones to achieve more important goals. To begin with, you can't develop many graduate students unless you can pay their assistantships. In some academic circles, it is considered gauche and uncomfortable to discuss the thousand-pound gorilla in the room, namely the firm (sometimes unspoken) requirement that young professors need to become effective/successful proposal-writers.

Why folks should be embarrassed to discuss this issue openly may appear somewhat strange. It has multiple causes, perhaps the most fundamental one, it stems from the utopian impulse that we would all prefer to emphasize scholarship and teaching over salesmanship, and avoid measuring research excellence by research funding.

While excellence of research is not deterministically measured by the volume of research funding, the two are obviously correlated. Another reason the gorilla in the room is frequently not discussed is the impulse many faculty have to "never let their colleagues see them sweat" – some may not discuss any difficulties that they recently had (or currently have) due to some misplaced notion that such openness about difficulty in securing a grant shows weakness or vulnerability. Well, virtually every professor has to "sweat" research sponsorship occasionally. Even great researchers and proposal-writers will not succeed all the time, and the larger the projects, the larger the gaps, dollar-wise, in making payroll when there is a budget fluctuation. So, if you have difficulties securing grants, you can be certain that this is not a unique experience and your colleagues are likely in the same "soup." These research funding headaches are almost universally shared by your colleagues, even if they are reluctant to openly speak about their problems!

The absence of straight talk on these issues can lead to blindness and a slow start by young professors struggling to understand the life they have chosen. Here I adopt a pragmatic perspective that is readily appreciated by your department head. It is obvious that the payroll of the department must be met if the department is to exist. Consider the economics of higher education from the perspective of a department head.

It is vital that aspiring faculty understand the broad outline of the "business model" of a successful academic department. What is the source of the money needed to operate the department? At virtually all public universities, the "hard dollars" associated with the combination of state funding, tuition, and other recurring monies are inadequate to run a strong undergraduate program. It would most certainly be impossible to operate a strong graduate program without additional financial support. Based on my direct and indirect knowledge of top engineering departments' budgets, including those whose promotion data are given in Tables 1 and 2, I note that between two and four "soft" dollars are annually raised by the faculty for each "hard" dollar available to the department.[11]   The exact ratio fluctuates by institution and by year within a given department. Furthermore, the "hard" dollars cannot simply be taken as a given fixed budget, because they are generally coupled through some explicit or judgment process to the size and quality of the graduate program, which again depend to a significant degree on some running average of soft dollars generated by the department's faculty.

Put in concrete terms, consider the following typical example (2012 dollars): Graduate students are usually funded mostly on soft dollars, so 150 funded graduate students @ ~$50K/yr each (all costs included) equals ~$7.5M of soft money.   Of course, some of the students will be funded on fellowships and teaching assistantships, but at least $4M would be needed to cover over half of the 150 funded graduate students. With the hard dollars available totaling about $4M, (as a typical example, note this may not fully cover the academic year faculty salaries for a 30 Full Time Equivalent (FTE) faculty department), we begin with the department being several million dollars in the red, and we have not yet discussed funds needed for computers, laboratory equipment, post-doctoral researchers, technicians clerical/ administrative staff, un-sponsored travel and faculty summer

---

[11]"Hard" dollars refers to the portion of the departmental budget from recurring, relatively firm sources (state outlays, tuition, gifts endowment earnings). The portion of the budget that must be won annually through the successful faculty research proposals is widely called "soft" money. Most faculty members would assert that their "soft" research money is actually "hard" to win!

salaries. In this example, the size of the graduate student body divided by the number of faculty is 150/30 = 5, which is typical for a top-tier program. Thus the economic picture is easy to scale, and I find that it is relatively invariant from the point of view of the individual faculty member. You should be getting the picture. Were all faculty of any research active department to suddenly stop raising soft money, it would have effectively put up a "going out of business sign" with regard to its graduate program.

A reasonable degree of financial stability, beyond faculty competence, is the most fundamental necessary condition for long-term success of the department. As a consequence, a strong department having 30 tenure-track faculty will generate over 100 new research proposals each year, with typically about half of them being funded. A nationally competitive engineering department cannot run a strong graduate program without winning, on the average, several million dollars annually for new research projects. The average research expenditure of the departments considered in this study was over $12M/yr.

From any rational evaluation of this financial situation, it follows that long term stability and health of a nationally ranked department dictate that the grant-writing activities be shared by the faculty broadly. Few engineering departments would knowingly offer tenure to young professors judged to be a poor risk for attracting external support for their research, because of the dependence of the department, especially the graduate program, on the success of faculty research proposals.

Certainly, we should expect a time-varying degree of success in proposals submitted by any professor, so the standard for young faculty must be flexible to reflect this truth. The probationary tenure-track period provides an adequate period of time for a young professor to establish some clear evidence that he/she can write strong proposals. Furthermore, some degree of early success obtaining research sponsorship is a *de facto* requirement for promotion and tenure. Put another way, as badly as the senior faculty may want to add the academic strengths of a young professor, they typically would not volunteer to indefinitely support the new faculty member and her/his graduate students. Convincing evidence that the young professors can "pull their weight" by teaching well, researching well, publishing well, *and* writing a few winning proposals are widely seen as reasonable rites of passage.

However, most engineering professors I know, privately (and publicly, actually!) criticize our current system because it exerts excessive pressure on the faculty and on young faculty, in particular, to write winning proposals. Obviously writing proposals consumes significant blocks of time, and lack of success can certainly lead to frustration. Proposal writing inherently competes for the time and energy needed for teaching and scholarship. That being said, we must eventually come back to square one and face the present day reality: The revenue stream needed to re-structure engineering education at public universities is not there. The fraction of state budgets that would be required to get back to the "good old days" of the 1970s (when ~80% of public university budgets came from state dollars, tuition and endowments) is a fantasy-land dream that "ain't gonna happen" anytime soon.

As I look across the nation and extrapolating the evident trends, direct state support for public institutions will soon be less than 10% of the total budget in most states. By 2020, it is apparent that most state universities will be "state affiliated" universities as measured by fractional contribution of the state to the total institutional budget. The faculty generated sponsored research programs are already the dominant source of the annual budget at most Tier I public universities.

Regardless of how the system has evolved to its present state (and that would be another interesting study), or how we hope that it may evolve in the future, a young professor time-sharing a personal crusade to locally change the system, so he or she does not have to be so "entrepreneurial," would most likely not be choosing an effective approach to career strategic planning! The present system, with whatever flaws, works reasonably well and is routinely navigated by succeeding generations of people entering the professorate and moving through the ranks. Since there are many productive and happy people in academia, how hard can it be?

If one reflects on the positive aspects of the necessity that some of our research must have "relevance" (which is an academic code word that some agency or industry deems it of sufficient interest to fund the research) then there are some significant positive outcomes. For one thing, implicitly requiring professors to engage in sponsored research is a strong antidote that prevents professors from resting on early laurels and "re-doing or embroidering" their dissertations throughout their careers. A fresh PhD in aeronautical engineering during the mid-1940s would likely have been an expert in high performance propeller-driven aircraft, and unless he or she remained an aggressive researcher over the subsequent three decades, there would be no way this person could have participated in the explosion of jet and rocket technology during the 50s, 60s and 70s. The same situation exists, to varying degrees, over virtually any three decade interval and in any modern engineering discipline, because the half-life of your PhD level education spans less than one-third of your anticipated career.

In short, engineering is at its heart a change-driven enterprise, so strong incentives for faculty to continue learning and expanding their spheres of competence (thus retaining their ability to address currently important problems over their multi-decade careers) are obviously healthy. Indeed most top departments want their strongest faculty to *lead* the evolution of their fields. It follows, if they are leaders, that finding research support will not prove so difficult. So another way of looking at conducting sponsored research (along with perpetual preliminary research, proposal-writing, technical marketing and so on) is to recognize that the most important long term side benefit is enhancement of faculty competence and relevance. A second exceedingly important benefit, the research-active faculty member will have many contacts in the government, industry and other universities, so placement of students in good positions after graduation is vastly improved.

To return to the main thread, the majority of eventually successful young professors that I know have won some research grants by the end of their third year. This rather short time to write some winning proposals (while also teaching and publishing effectively) is, from some viewpoints, remarkable. This efficient transition is possible because most engineering programs have developed a system of benevolent mentoring designed to help newly minted professors successfully meet the challenges they face. Furthermore, the competition to select assistant professors leads to a group of extremely promising and highly motivated people that are made of the "right stuff" to perform cutting edge research.

Most departments provide significant financial support in a startup package that effectively endows young professors with funds to support one, two or more students, to build laboratories, to support summer salary, and reduced teaching loads for a few semesters. From a strategic investment viewpoint, the department and the university consider these startup packages, and indeed all tenure-track faculty appointments, as "academic venture capital." It should be obvious that literally everyone has a strong incentive to see these investments succeed, so we should expect a supportive environment where young faculty members typically receive benevolent mentoring from their senior peers, especially during the first few years of their career.

To get an idea about "how much of the *M word* are we talking about," we can use the $50K/yr number per graduate student, and take into account that in steady state you will need about 4 graduate students funded, and a few undergraduate assistants. Additional funds will be required to cover your summer salary, laboratories, publications, and travel. You can do the math, it is evident that stability begins to set in around two or three grants totaling more than $200K/yr, and this is why annual research support per faculty amongst top 20 programs is usually greater than $350K/yr. All this research support discussion is in 2012 dollars, so adjustments may be required to extrapolate these remarks to the future. The startup transient is typically three to five years. Experimental researchers may require significantly more support than researchers who do only analysis and computation. There is not a "one size fits all" expectation with regard to startup packages and success in proposal-writing.

As a suggestion for a five year goal, an assistant professor should strive to take the lead in securing three to five funded projects with the total funds raised in excess of $400K (as a nominal target, during the first five years). More is likely better, and less will frequently prove good enough. However, securing and managing more than about three times that amount will most likely challenge your ability to find time for strong scholarship and teaching. Obviously these gross guidelines will vary by institution and local circumstances. The expectations for promotion to full professor are substantially higher; Associate professors are usually acquiring more than $300K/yr and supporting about five graduate students at the time of their promotion.

So my message to young assistant professors regarding the M word is the following. You should recognize that raising a certain level of extramural dollars is a necessary step in learning how to fulfill a long term requirement for being a successful engineering professor. Accept this truth and learn how to achieve the balance needed to succeed in the full spectrum of activities (teaching, scholarship, service, and proposal-writing).

Given that you achieve at least a minimal level of financial support, *what you do with funds you have in hand is an order of magnitude more important than the volume of financial support obtained.* Your senior colleagues understand that the long term quality and reputation of a department are built upon the quality of the results you achieve (students, papers, presentations, texts, laboratories developed, and favorable visibility). All meaningful measures of excellence are correlated to volume of financial support for a department. However, the excellence of an individual's research program is not determined by the size of his or her research budget. So let me cut to the chase: *A hypothetical Assistant Professor that raises a million dollars a year for five years – but does not generate high quality scholarship, does not teach well, and does not mentor graduate students well, will definitely not get my vote for promotion and tenure.*

A department, college and/or university will usually get an abundance of what it rewards. If the faculty is composed of excellent technical marketers, who are rewarded mainly for raising money, but who are poor academics – well, we all can extrapolate where that leads. It certainly does not lead to a top quality academic program. Using a "grading" analogy, your performance in winning research sponsorship is graded "pass/fail," and must be a pass. Your final "grade point average" in the tenure and promotion quest is based on judgments about your academic excellence (mainly, your scholarship, teaching and mentoring). A "pass" grade in level of research sponsorship success simply means that you have made yourself affordable, and given that this question is answered affirmatively, your senior colleagues will be far more concerned about the real issue, the quality of your work. These remarks idealize reality, but I think this perspective is the correct one.

Beyond doing good research, I offer a few practical suggestions to you (my young colleagues who are seeking to succeed in grantsmanship) regarding finding your fair share of research funding. Begin by understanding your potential funding agencies, and especially, focus on which of the "hot" current and emerging areas of emphasis are most nearly aligned with your set of competencies and technical ambitions. Establish a "top ten" list of leading colleagues well-placed to inform your research and proposals. A good mix would have about half of these people in academia and the other half in government agencies and industry. You should make it an important part of your strategic plan to know these people well, their technical contributions and their opinions of the important technical challenges in your field.

You should also find ways to interact frequently with your top ten colleagues about their ideas, especially the connections between your ideas and theirs. If this colleague is in a potential government laboratory sponsor or is a potential industrial partner, you should have these interactions in advance of submitting proposals. If a potential sponsor's only time spent with you is when he or she is a target of aggressive salesmanship, you may not find a welcome mat for your visits, emails or telephone inquiries. Central to your success with research proposals is investing the time and energy to become well-acquainted with peers in your field and gain their respect and trust. Getting past the "who is the author of this proposal" reaction with which sponsors and reviewers greet proposals from people they don't know, is a vitally important part of your career development. They need to learn that you bring "something significant technically to the ball game" and that your words are backed by effort and significant results, *before* they are asked to invest their funds.

Young professors have a secret advantage. Their more seasoned colleagues and potential sponsors are always looking to connect with the next generation of rising stars – so set your goal to be one! You will likely be surprised at how welcoming the climate is for well-prepared individuals who have thoroughly done their homework before they get their first turn at bat.

There are a number of strategies you might employ. First and foremost, read a lot and prepare for each contact with a potential sponsor or collaborator. You should find approaches that result in you spending significant time in the facilities of colleagues and interacting with them one-on-one. Recognize that implying to a technical colleague or potential sponsor that his/her thinking about a problem is dumb or out of date, in light of your revolutionary new approach, may not be the most diplomatic or effective way to pursue research support. Find ways to let your work speak without heavy-handed, derogatory language on potentially competing ideas. I recommend that you treat technical colleagues and potential sponsors as respected peers and potential collaborators. After hearing their ideas, studying their work, reviewing in detail their existing research projects, and understanding their goals and approaches, then come back to them with refinements of your ideas motivated by, and clearly connected to what you have learned. Where possible and relevant, use and generously reference competing ideas or methods. After several iterations of technical discussions with a potential sponsor, when there is perceived to be an opportunity for your research approach, *then* strategize with that sponsor on writing a proposal.

It is important to understand that the basic funding agencies in DoD (AFOSR, ONR and ARO) are "research service agencies" for, in the case of the Air Force Office of Scientific Research (AFOSR), the ten laboratory research directorates of the Air Force Research Laboratory (AFRL). So, to be successful in AFOSR, you need to know at least some of your most relevant technical colleagues in AFRL who are nominally served by the research funded by AFOSR. You should also meet the relevant program managers at AFOSR. If you have found a strong interest match with AFRL colleagues and understand their perspectives on the challenges in your field, before you visit the AFOSR program managers, your white paper/proposal writing efforts will have a greater likelihood of success. If (and frequently, when) your initial white paper or proposal is not met with the level of enthusiasm that will lead to funding, listen carefully to review criticisms and clues to variations of your approach or related problems that might fit into the program manager's set of fundable research thrusts. Also listen to insights that may help you learn more about competing areas or approaches that are related to your proposal.

If (or, I should say, when) you have a proposal turned down, and even if you strongly disagree with a review assessment, use calm logical discussions to explore in detail the perceived shortcomings of your proposal. In the event that there is an opportunity to submit a revised proposal, be sure you respond thoughtfully with revisions that make it clear that you heard and acted on the review inputs. Avoid behavior that might be perceived as you being paranoid or thin-skinned. Be persistent, but always invest significant effort to address any issues "left on the table" between interactions with the program manager. There is a thin line between being "purposefully persistent" and being a pest, so try not to cross it. Always base your interactions on fresh results since the last conversation, so you are not perceived as rehashing previous arguments. When you sense the program manager has heard enough and is not going to change his or her mind, don't keep pushing the point. If possible, try to end each interaction with a clear plan for your action. Come back only when you have "something different" likely to be seen as significant to show, and begin each subsequent discussion with a clear report on what is new.

Sometimes several trials are required, especially for the first successful proposal. Multiple occasions will arise in your career, however, no matter how just your cause or how hard you try, that you will not be able to convince a program manager who has made up his or her mind negatively about your research proposal, or worse, on your potential to contribute in the area he or she has control over. On these occasions, you must be tough, and avoid angry outbursts that accomplish nothing. I speak with authority because I have been there and done that. On the other hand, I have lived to see some of these folks dramatically change their opinions in a year or two and see (whether or not they explicitly admit it!) when subsequent successes make them realize that they missed an opportunity by not sponsoring your work at an earlier stage. So, the best revenge for any perceived injustice is go have a good life! Time well-spent will usually bring justice to these short-term setbacks.

My experience indicates that the time spent in trying hard to address review criticisms is almost invariably productive, in the sense that the you will better "own" competing research threads, understand the problems of interest to the funding agency, and you can almost certainly give a more compelling, complete and in-depth presentation when the next version of the proposal in question (or a related proposal) is submitted. Furthermore, I have found that a significant fraction of these good faith iterations eventually converge, if you are persistent while keeping your cool. These transactions, even when they do not result in a win, can earn the program manager's respect. It may mean that your next proposal gets the program manager to reflect on the truth that you do not go quietly into the night just because you get a negative review. Frequently after you win a "proposal beauty contest," you will feel (correctly!) that a significant fraction of the proposed research had to be completed *before* you could get the project funded! You should always view a successful proposal not as an "award", but rather as a "reward" for having done your homework that was the foundation for the successful proposal.

It is vitally important to understand that finding financial support for strong students is frequently an easier sell than finding support based on the merit of your ideas! While this may be tough on your ego, this truth will often present an opportunity, if so, take advantage of it. It is very important that your potential governmental and industrial partners see you and your students as a pipeline of well-educated, highly motivated individuals who can contribute significantly to *their* current and future success. As is brought into clear focus in the 2007 National Academies report *Rising Above the Gathering Storm* [10], there are disturbing global trends in the US technical "labor pool", with important implications for our future.

Most major research and development organizations are keenly aware of the need to improve the pool of well-educated engineers, scientists and researchers for the next generation workforce and are receptive to teaming with universities that are excellent talent pipelines. It follows that highlighting important contributions or strengths of your students, especially those nearing graduation, will be received favorably by many potential employers. Their interest in your students creates a favorable pressure gradient that motivates them to find common ground with you and your research interests. Proposing research projects which involve internships by you and your students in the facilities of the sponsoring organization is therefore likely to be warmly received.

Altered strategies are needed when writing proposals to the National Science Foundation (NSF). NSF does not usually conduct or sponsor research with priorities established by a network of strongly coupled applications-oriented laboratories. NSF basic research can and should be less coupled to near-term applications. This truth is both an advantage and a disadvantage as regards young professors breaking through to achieve early success. The basic research agencies being coupled with more applied research laboratories, as is the usual DoD model, provides corporate memory to build upon your successful basic research, and allows a natural path for the agencies to mature basic research to contribute to advanced engineering practice. Also, the above incentives for identifying and developing the next generation engineering workforce are motivating for the program managers in DoD laboratories who are looking to hire stellar talent, but this is typically less so for program managers in basic research funding agencies such as NSF.

The Career proposals should be pursued aggressively by young faculty during the first two or three years, since these opportunities are open only to young professors initiating their academic careers. The Career proposal should usually build upon, but also "put significant technical space" between, the young professor's PhD dissertation research and the Career proposal. Early results along a research path to the future that goes in other directions beyond the PhD dissertation need to be established through your preliminary research (beyond your dissertation), and visits to speak to the appropriate NSF program manager should be done to present a white paper outline or draft of the proposal before generating the final proposal. This white paper, and especially the full proposal, should be vetted rigorously with benevolent colleagues who have a record of successful proposals to NSF.

In addition to the single Principal Investigator (PI) Career proposals, collaborative, multi-PI proposals are also encouraged. There are a number of opportunities each year with all of the above funding agencies, including NSF. It is important to recognize that early collaborative efforts for younger professors will likely be teaming with one or more senior colleague's projects as a co-investigator. Be aware of pitfalls. I suggest that you look for senior collaborators who are strongly interested in your career development, not merely in taking advantage of you as relatively inexpensive labor.

An important pitfall that I have observed that causes frequent casualities among engineering researchers is the combination of technical arrogance and technical ignorance. Ignorance is distinct from stupidity. Read the literature and initial ignorance will usually be curable. There is no known cure for stupidity. Also, even though it is natural to be proud of your ideas, do your best not to over-hype your ideas and make claims that any competent reviewer will see as you just "blowing smoke."

Frequently we engineers are guilty of some combination of myopia, ignorance, and arrogance wherein we develop a near-religious fervor that our approach to solving problem X is "the way, the truth, and the light," and we may project an undue, irrational intolerance of competing methods or criticisms. The worst mistake is when our preference for our ideas is built on ignorance of competing approaches. This is especially difficult and damaging when your ideas actually are the "best" approach, but you write in n arrogant style that rudely dismisses (or fails to mention) competing ideas. This behavior is a likely path to proposal rejection. You must project knowledge of the literature and appreciation of competing approaches. It is vital that we balance passion for our ideas and approaches with a broad appreciation for the most relevant set of competing ideas and literature dealing with the problem at hand. We should always strive to fairly discuss and compare the literature on competing ideas with the ones being proposed. In special circumstances where some other method is especially well-suited, make generous comments to recognize this assessment. Do not make biased or unsubstantiated claims. Obviously, if your ideas represent the new kids on the block, many of your reviewers will be proponents of the competing ideas you are seeking to "improve upon." Thus a delicate balance in your discussion of the state of the art and your new research directions is required to avoid un-necessary bruises and, especially, errors of omission. When the program manager has a perception that your analysis of the literature is fair and complete, and that your research ideas are both significant and presented in a context where the anticipated advance can be readily documented, then sponsorship of your research is much more likely.

I punctuate the above advice with true anecdotes from my personal experience. My first six proposals were rejected before I came to understand the several things I was doing wrong. Most of the mistakes I was making were relatively easy to fix, once I "got my mind right."

One key mistake I made leads me to give you some important advice: You need to be certain that you are targeting your proposals clearly on problems known to be of interest to the prospective sponsor. You may have thought the utility of your ideas was obvious, but you should never fail to state important features and their relevance to your sponsor's specific goals. Think very carefully about your proposal from the perspective of your reviewers and potential sponsor. "Cover the waterfront" in discussing the literature. Get second opinions from colleagues on your draft proposal and respond to their constructive criticisms.

I raise another issue for those engineers among us who consider ourselves really strong at mathematics of engineering science.[12] I learned a painful lesson about the level of abstraction, as I mentioned earlier. We can all easily "snow" our readers with a treatise of abstract mathematics and specialized language that

---

[12] Many engineers mistakenly feel the intellectual high ground is achieved if they are able to raise the level of abstraction of their technical discussions to the point that only stellar mathematicians can read their work. My experience indicates this is not really the high ground. I believe that Gibbs and von Kármán were both far stronger mathematicians than any engineering professor I have met personally in my four decades, and these two superstars communicated their sophisticated ideas clearly to the broad engineering audiences. Most engineers who fancy themselves as mathematicians need to spend a little time in a mathematics department where humility will usually come soon enough. The key is to have something new and interesting to communicate in your proposals and papers, and in many cases, overly abstract presentations simply obscure the message you need to deliver clearly. *Elegance does not equal abstraction.*

only the most well-informed in our area of expertise has any hope of tracking. Using an overly abstract presentation of your approach that requires the reviewers to work harder than necessary to understand your ideas will usually earn you bad reviews. Ignoring, or failing to give adequate discussion of, the challenges posed in the call for proposals and how your approach specifically addresses the challenges is usually fatal. Areas where you are not confident in the outcome should be identified as concerns or open questions, you can discuss plans for research that should help you find answers or solutions to the open questions.

Reviewers are most often busy, un-paid volunteers, so you should make their life easy by writing in a clear, well-organized fashion. Point out the key ideas, significance, and impact of your proposal. Try to make your writing interesting. Mathematically rigorous presentations have their place, obviously, but make sure your proposal will be fun to read, if possible. Ten pages of abstract developments without making physically interesting conclusions will typically not be warmly received. Also, use heuristic or qualitative insight where appropriate, perhaps in interpreting the meaning or utility of more abstract developments. If the theory has shown an approximate match with experiments, then by all means show these results, and discuss discrepancies as areas for investigation.

If you are targeting a proposal to a laboratory where there are potential government research collaborators as well as sponsors, ask one of them to review an informal draft submission before you submit the proposal formally. If this informal review generates a suggestion you can incorporate in the final proposal, then, certainly, do it if possible. Make certain that you refer to all pertinent work presently sponsored by the laboratory, especially if some of this current or recent work is being conducted "in-house" in the agency where you are seeking support. If competing work to your proposal has already been funded, define a promising new direction, perspective or application so that it is easy to see your proposal as exploring unique and interesting ideas.

Write in a way that generates a strong basis for optimism that you are serious about achieving significant results; pointing to historical examples if you have related accomplishments. Your proposed research needs to be viewed as helping make the sponsor successful, so understand his or her goals and how your proposal "fits in." I mention as an aside that learning and applying the lessons above resulted in all six sets of ideas embedded in my initially rejected proposals' being incorporated into winning proposals, so you should never, ever give up on your good ideas!

In addition to the above strategies for improving your ability to acquire sponsorship, I would add, especially to associate professors, that you should seek to participate in nationally visible professional service, for example on committees of the National Research Council, or on Technical Committees of your Professional Society. Curiously, such professional service can directly and indirectly make a constructive impact on your success in acquiring research sponsorship. You will have more opportunities to see what is happening in your field and to earn the respect of your peers through service functions. These service activities will frequently impact favorably your ability to conceive of relevant research and write successful research proposals, and also create teaming opportunities with colleagues.

One of my colleagues who researches at a national laboratory read a draft of this text and was amazed that faculty conduct research, and perform professional service, while raising, "on the side" over the half of the University budget to fund research. Furthermore, he was even more amazed that the faculty effort required to win 100s of research proposals per year is not explicitly compensated. National labs and industrial research organizations have "bid and propose" budget categories built into their overhead rates to cover the cost of competing for research. Universities implicitly depend on the faculty passion to generate this effort to effectively compete for research, on overtime. Most Tier I institutions have the found ways to distribute assigned teaching to provide lower classroom teaching assignments to partially accommodate effort to direct larger research projects as well as compete for next year's research budget.

Chapter Six

Good Teaching:
Developing People Lies Near the Heart of a Great University

First of all, if you do not enjoy teaching, and more fundamentally, if you do not feel you will take great pride in fostering the development of the engineers of the future, then the University is not a good place to pursue your career. Research is important, but coming to a university to pursue research without having a strong commitment to developing young people is a recipe for frustration and likely failure.

Researchers not interested in teaching and mentoring should pursue a career in industry or at a national laboratory. Students should not be viewed just as a resource (i.e., cheap labor) to be consumed in pursuit of a professor's personal goals. The notion that research is valued more than teaching and mentoring in the modern University should be roundly rejected and tamped down by the faculty and administration. Teaching and research are coupled and very strongly at the graduate level. Any university research project that does not foster the development of students should be rejected as inappropriate, in my opinion.

Certainly the modern University does not uniformly function with this Utopian set of values, and virtually every University will generate some mixed signals to aspiring faculty. The points of view I express are, I feel, the right perspective for young professors to adopt and are healthy for the senior professors to consider when mentoring young colleagues and to improve their institution. This perspective places the top priority on developing people; it is definitely consistent with excelling at research, especially when taking the student-professor teaming approach to scholarship. I believe that it is vital that you come to truly value your students and take pride in their progress. If you desire to maximize your career impact as a professor, then you should invest your very best effort in the development of your students – and your contributions in teaching and research will live forever through a school of thought embodied in your technical descendants.

So, if a selfish motivation is needed, placing heavy emphasis on mentoring and development of your students through involvement in your research is definitely the path to maximize your career impact. Furthermore, when you reach the prime of life (that magical period between sixty and ninety), a career of effective teaching and mentoring will almost always result in a large cadre of younger, more energetic, and hopefully benevolent technical collaborators who can help you maintain a small positive slope for your golden years' learning curve! I am really beginning to appreciate this benefit myself.

I might add that an "extended family" of former students can become in all respects a genuine family. A lifetime of investing in people has many qualitative professional and social rewards that cannot be quantified. When I reflect on my career, there is no doubt that my main impact has been a family of technical descendants whose integrated productivity dwarfs my personal career contributions by two orders of magnitude. I take great vicarious pride in their accomplishments as if they were my biological sons and daughters. In the next chapter, I give you suggestions about how to attract and develop talented graduate students. This may be the most important part of my advice.

Here are some suggestions and observations that I recommend you pursue to become an effective teacher:

- Avoid behavior that can be interpreted as snobbery or intellectual intimidation. Respond thoughtfully and respectfully to questions.

- Make it evident that you take pride in helping students acquire knowledge that puts powerful "tools" in their hands. Make it clear how the best methodologies apply to large families of problems.

- Be friendly and benevolent, but firm. Earn a reputation for fairness, but not as a "pushover" who, for example, revises fairly assigned grades when aggressively confronted by either complaints or sob stories.

- Project your love for and excitement about the subject matter in your lectures and interactions – this is extremely important and still compatible with your role as an effective "task master."

- During the first few lectures of each course, present basic material and reviews in a tutorial fashion that establishes goals and basic notations for the class. A quick review highlighting most relevant background material should always be included to begin each lecture.

- Find ways to use the first few lectures to evaluate the initial knowledge and competency distribution of the class. One useful device is to give a questionnaire with a list of terms or concepts; ask the students to tell you their state of prior knowledge on a scale of "never heard of it" to "understand it well".

- Give very specific remedial suggestions to those students with weak preparation, and if the student is a graduate student, have a conference with the students' advisor to help direct the remedial study and to motivate the students' best effort.

- Attend lectures of colleagues known as excellent teachers – take notes on their style and draw lessons to incorporate in your teaching. This can be very important during the first few years of your career.

- Invite colleagues to attend your classes and ask for constructive suggestions. Listen to what your colleagues say and respond constructively.

- Engage your students interactively. At least once per lecture, ask the class a thought provoking question that relates to "what is to come" later in the lecture and the class, and spend a few minutes discussing their answers.

- Hold realistically high standards.

The above thoughts apply to both graduate and undergraduate courses. For undergraduate courses, additional points are worth emphasizing:

- Regarding balance of "Theory versus Examples" of your lectures: Make sure your lectures on methodology are clearly related to problem solving and illustrated by multiple examples. In some lectures, you may find a way to start by discussing a problem and while outlining the kind of solution that is desired, raise conceptual challenges or "wish lists" for the solution that you can subsequently address in your lecture. Return to the problem and illustrate how the more abstract material just presented really helps to solve the problem originally posed.

- Do not forget to require students to show frequent evidence of what they are learning, "paying some homework dues," and tracking your lectures. Call on at least one student toward the end of each lecture to "help you" on some aspect of the lecture or solving an example problem, and have a constructive interaction based on their response.

- Learn how to effectively use teaching assistants. Require your teaching assistants to attend your lectures, and make sure that they know how to solve homework problems. Make sure both you and your teaching assistants are available during scheduled office hours. Spot check a fraction of the teaching assistants' grading to give them feedback and make it clear you are monitoring their work.

- Take a long view; assume that you will likely be teaching each course more than once. Therefore you should maintain systematic notes, collections of problems for homework and exams, in-class experiments, and so on; your early investments will amortize over several years of increasingly effective teaching and with less "ground breaking" work per lecture.

- Being a professor implies that you need to develop depth in the subjects you teach and also develop perspective on why the subject is important in the "real world". Rote repetition of the text material is likely boring for you and via telepathy, also boring for the students. They can tell when you are really thinking and enjoying the material yourself. You can follow a text to a reasonable degree of approximation and give your own perspectives that augment the text significantly. *You should develop and profess your informed points of view, forcefully.* Your lectures will be more interesting and have a spontaneous feel that won't happen otherwise.

- Demonstration experiments are very useful to motivate students, many of whom need physical "real world" reminders of the relevance and importance of the subject under discussion. If good in-class demonstration kits are available for the course you are teaching, acquire and use them, if not, arrange for laboratory tours where relevant concepts are well illustrated by visually appealing experiments.

- If the course has a laboratory on other "experiential learning" activity by the students, be sure to interact with the students before, during, and after the experiments to connect these experiments with the more formal lecture material.

- Finally, the most important advice beyond being well prepared: Project heartfelt enthusiasm and excitement for the material you are teaching, and stress the opportunity that this course affords for your students to efficiently acquire important, really useful ideas. Have fun in the classroom (not at the expense of your teaching, obviously); just show that you enjoy the material being presented and more importantly, you enjoy interaction with students. *If you think back on your education, your best teachers projected excitement about the subject and showed a strong interest in people, you in particular.*

Chapter 7

Mentoring Strong Graduate Students

Assuming you are pursuing tenure in a department that offers graduate degrees, your goals should definitely include developing exceptional MS and PhD graduates. There is no substitute for starting with great "raw material." Recruiting well-qualified graduate students and finding those that have strong motivation to work in your areas of interest are the most fundamental steps. Once they are on-board, you and your students should develop a partnership with *the success of the partnership measured by how well the students progress.*

As the senior partner, your central task is to lead and help the student to understand the research goals. Also you should lead in establishing the initial directions of the research, in pursuit of the goals. Thereafter, you need to give frequent supervision and feedback on progress, and share with the student your personal research results in parallel with their effort. You should also help the student to digest the key literature and suggest remedial studies to better prepare him or her to pursue the research.

Here are some specific suggestions on how to find and develop good graduate students:

- Do not just depend on students to "show up" unattached in each annual recruiting class. While some excellent students may indeed be in the unattached category in each recruiting class, there are effective, *active approaches* to finding "your own annual crop of students." I believe that the key to success in consistently finding good students is for you to personally recruit graduate students as a year-round "background job." First and foremost, recognize that you only need to find one or two strong new students each year, so you can be patient and evaluate multiple candidates. Some suggestions to enhance your ability to locate excellent students:

- o Communicate with colleagues around the world to let them know that you are looking for strong students.
- o Give frequent seminars at peer universities and keep your eye out for student prospects.
- o Every time you recruit a great student, ask that student to help you identify and recruit other superstars from their previous institution.
- o Respond warmly to prospective student inquiries with personal notes, links to web sites, papers, encouragement, and frequent follow-ups to stay in touch, especially with outstanding prospects.
- o Remember the kind of students you are looking for will have many options – if you treat a stellar prospect rudely or ignore their inquiries, you likely won't hear from this student again.
- o Teach effectively at the undergraduate level and cultivate the interest of promising students. Involve strong graduate student prospects in your research while they are undergraduates.

- Treat graduate students you recruit "like he/she has potential to be a superstar," but make it clear that to realize their potential, they have to develop passion and invest in their future with high quality effort. If you are not highly optimistic about a prospective student, look for another one.

- Work with each student to establish clear goals for one to three month intervals. Frequent meetings are needed:

○ Give clear advice when you meet with the student, but leave room for their initiative and ideas. Make sure that they know that you expect to hear their ideas on alternative directions, especially after they have been working on a problem for some time.

○ Even if you are initially unexcited by their ideas, treat their ideas (on how to pursue a research goal, for example) with respect, but help bring objectivity by requiring them to do fair comparisons and contrasts with competing ideas.

○ The frequency of meeting with the students can be decided based upon how much contact you feel is required. Once per week with each student for about an hour, or twice for a half hour is recommended. Fewer than two meetings per month is too infrequent!

○ The first order of business when you meet should always be to ask the student to report what activities and progress he/she has made since the last meeting. Negligible progress twice in a row, and especially negligible evidence of effort, is unacceptable. However, activity cannot be substituted for progress. Use judgment before "coming down hard", but let them know that you are very interested in seeing some progress, and soon - let them know clearly when you are not satisfied with their effort or progress, but avoid explicit or implicit "I will fire you" or other "threat" messages. *Whenever possible, depend on the power of positive thinking and cultivating passion about solving the problem at hand to motivate students that need to accelerate.*

- Meet with your students in a group about once per month with students taking turns presenting results and responding to questions in front of their peers. Don't let students get away with "snowjobbery or vacuous fluff" in their presentations - find ways to steer them toward presenting with technical clarity and integrity. Praise progress in front of peer students to implicitly set examples of the quality of effort and thinking you expect. Constructively criticize poor progress and share specific suggestions along with your expectation that the student will soon have good progress to report.

- Give students responsibility for drafting conference and journal manuscripts, if not their first co-authored paper, then for sure the second. Work hard to constructively criticize their organization and presentation of ideas, as well as their writing style. Be generous with authorship – if a student contributes, be sure their contributions are generously reflected in the authorship. In my opinion, papers from PhD dissertation research should nominally have the student as first author. Reward excellent performance by letting students give some papers at conferences.

- When your research sponsors visit, don't do all of the presentations yourself – put your students on stage to present some of their contributions. Obviously, for a high stakes sponsor visit, you will dry-run all presentations.

- When a mentored-to-completion student graduates, especially a PhD student, help them with the job search process, including improving their resume, any presentation materials they are using for their job interview, and providing them with as much insight as you can as regards the organization they are considering. Help them prepare for the "tough questions" that frequently occur in interviews by raising these kinds of questions yourself and ask them to defend their work, concentrating on aspects that could be perceived as weakest, most incomplete, or in aspects where competing ideas are evident in the existing literature.

- Have a party for each successful dissertation defense, announcement of a best paper award, fellowship award, and so on. Celebrate the student successes and visibly take pride in their significant contributions in front of their peers.

Chapter Eight
The Four-To-The-Fourth Rule Of Thumb

So how do you set short term goals and balance your time to maximize your chance of success, given the simultaneous challenges of teaching, mentoring graduate students, research, and successful proposal writing? The first thing I emphasize when I begin mentoring a young professor is to recognize the very strong coupling of these functions and to take full advantage of this truth. You should fold the research results of your draft conference and journal manuscripts into research proposals, involve your graduate students in research and paper writing, introduce some research results into "motivating digressions" during your graduate course lectures, or use some of your research results to develop mainstream material for a special topic course, and so on. Regarding annual goal-setting for the volume of your output, I usually simplify my advice to young colleagues as the four-to-the-fourth (4^{4th}) Rule of Thumb:

- *Recruit and find ways to support 4 or more excellent graduate students working hard under your direction.*

- *Write and submit 4 excellent journal papers each year.* While you cannot directly control the rate of acceptance, the surest path to guarantee no journal publications is to submit none, or to submit papers of low quality. As is evident in Table 1, the average rate of journal publication exceeds two journal papers per year for those gaining tenure and promotion. *While the goal is to write about four good papers each year, submitting one or two good ones is by far preferable to submitting four weak papers.*

The rate at which the work will mature varies, so two good papers submitted might be the right answer some years, and six might be the right answer the next.

- ***Write and submit 4 or more strong research proposals each year.*** Never, ever give up on a good idea – if a proposal gets rejected, respond constructively to criticism and re-submit, incorporating changes and newer results. If reviewers are "missing the point" of your work, find clearer, more creative, and interesting ways to present your ideas and do more preliminary research to make it attract the positive attention of the reviewers. Since next year's research support depends on this year's proposals, this axiom is very important: *You should have one or more solid proposals under evaluation at all times, even though your research may be presently well-funded.*

- ***Attend 4 or more national/international meetings each year and present one or more well written and delivered papers at each.*** You should bring your "A game" to the meetings every time and present your ideas and significant new results in ways that capture the attention of your audience.

In all research presentations (proposals, journal papers, conference papers, seminars), do your homework and give careful thought to comparisons and contrasts to other approaches that are fair. Avoid making claims you can't support and by all means, avoid making un-necessary bruises by making poorly thought through criticisms of other's work.

Following the 4^{4th} rule, together with quality teaching and research, is usually sufficient to set young professors on a high probability course to achieve tenure and promotion. For Associate Professors, I would add the need to engage in significant national professional service and also within your university, through leadership/service activities at the departmental, college, and university levels.

The above 4^{4th} rule of thumb should obviously not be taken as rigid law, but rather as an approximate volume benchmark to guide your thinking. The quality of your results is of course much more important than volume, however I am sure you know that significant volume of output every year is a reasonable expectation. Developing a strong goal-focused work habit should make appropriate volume and high quality of output happen and position you well for tenure and promotion.

You should temper my above advice by seeking benevolent counsel from senior colleagues and use your own local observations of the traditions in your department – so the $4^{4\ th}$ rule of thumb is simply be the starting point for establishing guidelines reflecting the expectations in your zip code.

Chapter Nine
Concluding Remarks

I suspect that we have all met competent professors who were not socially enjoyable company. On the other hand, some of my most respected and accomplished colleagues are at once highly competent and have a graceful way of interaction with friends and colleagues; they have mastered the fine art of being self-confident without being egotistical. As you seek to advance in your career, I hope you reflect on this observation. In particular, you can definitely gain respect while resisting the temptation to blow your own horn often; try to let your competence and hard work speak for itself. I also remark briefly on a less offensive social habit of us engineers (I must confess that I have had problems with this habit throughout my career[13,14]) – as a group, I have observed that we have a tendency to over-explain our thinking on whatever subject is under discussion. Too much information (TMI) on most any subject can have a reverse-effect and turn off folks that you need to listen to your ideas.

---

[13] One night 25 years ago when my daughter Kathryn was in middle school, she walked quietly into our house and asked her mother if she would come out and show her Mars and Venus. Elouise (my wife) said, "Kathryn, there sits your Dad, he knows a lot about astronomy, why don't you ask him?" Kathryn turned to her mother with a cloudy expression in her eyes and replied "Mom, nobody wants to know that much about it!"

[14] Actually, as a postscript to the above footnote, I recently told this cute story to my esteemed friend and colleague Bill (W. S.) Saric, who said, and I quote, "Junkins, you didn't learn a damned thing, did you?!" Bill has always endeared himself to me with his special blend of caustic sweetness, I know he meant that little zinger in the most positive and constructive way I could possibly interpret it! I got even with him, but we'll just let that rest. You may have noticed by now – I try to have a little fun while I write (and similarly, in my social and professional interactions). I have found that being overly pompous and formal are not highly correlated to being professionally competent – there is an implicit lesson here … learn to truly have fun while you work and interact with students and colleagues, you will be a lot more fun to be around and this is definitely not inconsistent with being an accomplished academic.

I was deeply hurt by my own daughter implicitly accusing me of being long-winded and verbose, and now that you are nearing the end of this advice book, I am certain that you would agree with Kathryn and Bill Saric (☺) – I have really learned my TMI lessons well! On the other hand, I have developed an important personality trait that I can recommend that you cultivate: Learn to laugh at yourself and to make an occasional joke at your own expense. Sometimes an embarrassing situation is just an unplanned opportunity to show your ability to handle adversity.[15]

Of course, a modest degree of horn-tooting image projection and providing TMI are not fatal (otherwise, I would have been dead a long time ago), and the extremes can be tempered to socially acceptable bounds if you develop an ability to self-police your own "over-the-top" behavior. Some folks just naturally get more comfortable in their skin as they mature. I have known a few young professors who age like a fine wine – they may start off as bragging loudmouths and develop more pleasant personalities as their self-confidence improves.

While you can have a lot of fun with a career as a professor, ever so often, will encounter a moral, ethical or legal dilemma that is

---

[15] Here is a true story from the *Junkins Chronicles*: In my first interview for a faculty position, I was planning to use plastic transparencies (in a pre-powerpoint era, obviously) for my seminar and an impending disaster was brewing during the day prior to the hour of my presentation. I was traveling light with a combination briefcase and suitcase. In those years, I used "Ice-Blue Aqua Velva" aftershave lotion, and the bottle of aftershave lotion was busy leaking highly aromatic deep-blue liquid on my clothing and, horror of horrors, mostly all over my slides. When time came for my seminar, I retrieved my bag and opened it. A blue cloud emerged and the first three rows of the seminar room quickly evacuated to a safer standoff distance. To say I was embarrassed is an understatement. As I used some paper towels to clean enough blue gunk off my slides that I could use them in my presentation, I began making jokes (to keep from expressing how I really felt). As I was about half-way through the 5 minute cleanup operation, began to relax a bit and I remarked: "At the end of today's lecture, I will be happy to entertain your technical questions, however, I will rule out of order any comments about how this presentation smells!" The audience erupted in laughter; I relaxed and gave a good seminar. That evening at dinner, the Dean said: "Young man, I was not able to follow all of your presentation, but I must say, anyone who can recover from such an unmitigated disaster must be made of the right stuff! So if you remain interested in a position here, I will be delighted to extend an attractive offer." I took away a lesson, with a little spontaneous humor, you can survive anything!

definitely not fun, and you will almost certainly be tested in ways that, perhaps, no advice book can prepare you for. I have had a few of these challenges arise in my life, and on the average, there is no way to make them fun. When your time comes to stand up and be counted, find your courage and think carefully before your act. Try to find the most logical, effective and honorable paths to deal with the particular challenges you face. Act as quietly and privately as you feel is appropriate, but publicly if you feel you must. Acting with integrity that meets your standards will virtually always enhance your career, especially *measured by your career satisfaction*. Your satisfaction is after all the most important metric of success in your career, so don't you ever forget that!

In some cases, you may find that you are able to act with grace under fire that reveals an admirable fusion of courage, logic and courtly manners that may enhance your standing among your colleagues. In other cases, the dilemmas you face may not be visible to colleagues, but you will always know. Academics are pretty perceptive bunch, and since grand-standing image projection is easily seen through and typically back-fires, you should avoid such behavior. You should definitely not go shopping in a "morality gun-slinger mode" looking for opportunities to display self-righteous claims of courage, holier-than-thou morality (or for that matter, more perceptive-than-thou snobbery); an offensive showmanship legacy can undercut your credibility and handicap you when a truly serious crisis erupts. You need the "political capital" of trust in your judgment by your colleagues. Wasting this by frequently engaging in Don Quoite or Chicken Little crusades "flailing at windmills" or making exaggerated claims that "the sky is falling" in protest of minor concerns is a mistake – don't promote molehills into mountains.[16]

---

[16] If you don't know it already, you will soon learn that "faculty morale is always at an all-time low!" ... this cute quote is due to [Hicks, P. (2003) *Teachers: feeling the heat, Directions in Education*, Australian Council for Educational Leaders, vol. 12, no. 16], and it captures something important. Some faculty colleagues have a "happiness set point" of about -50 on a scale from -100 (morbidly depressed) to +100 (deliriously happy). It is easy to get swept into negative feedback loops that are much ado about very little. Choose your crusades carefully and filter input from "nabobs of negativity."

Some (hopefully, only a few) serious challenges will arise on their own, and as I said earlier, they typically are not fun. When a serious challenge involving you does arise, I hope you don't cut and run for the wrong exit, or make poor decisions, or engage in behavior that would not make your mother proud. I can look back and see a few defining moments where vitally important, high stakes issues were involved, and I had an opportunity to choose an "easy or politically expedient path" or the "most honorable path." I take pride in the way I handled most of these situations. However, in hindsight, I recall a few instances where I could improve on how I handled crises, and I would really like to have a do-over on these! While there are no do-overs in life, you can and should learn from both your mistakes and your triumphs. I have strived to learn whatever lessons were possible through constructive post-mortem evaluations, in the wake of each crisis resolution. Sometimes there will be an opportunity to talk through a pending decision with someone external to the crisis whose judgment you respect – I encourage you to seek wise counsel if time permits. Such consultations helped me immensely in one instance I recall. Through the lessons learned, I have grown and become better in dealing with the few painful challenges that have come my way to date. If you engage in frequent reflection and constructive self-evaluation throughout your life, you may indeed age like a fine wine!

It is impossible to write an advice-filled text such as this one without some parts coming off as "preachy thou-shalts and thou-shalt-nots." Given this un-avoidable truth, I trust that my advice will just be accepted in the same spirit that I have offered it – as my honest attempt to be helpful. I do believe most readers will be able to find some useful advice herein, but I am also confident no one will agree with 100% of my opinions on this subject, much to my delight!

In reviewing the final draft of this book, in my self-critical mode, I felt that I have consumed a lot of this book giving "navigation, guidance and control" advice to guide your thinking and steer your effort, but I have not done justice to some other issues that are really more fundamental: Namely, this text on "how to succeed" may not have adequately dealt with an even more important issue. I don't believe I have conveyed how much fun one can have as a professor! Let me address this important issue briefly by raising the following question: In what other profession can you work with bright young people and highly motivated colleagues, and where you have both the privilege and responsibility to invent over half of your job? What a job it is, surfing the curl near the leading edge of change in an exciting innovation-driven field?! I have been at it for four decades and can tell you that I am still having a great time.

With regard to the importance of this book, I have no illusions that it will be found indispensable. I will settle for "very useful!" Clearly, thousands of people have been routinely moving through the ranks of this evolving system nicely for decades, including me, without having read my opinions on the subject. However, I hope that I may have provoked some useful introspection by both young faculty and my more mature colleagues. I can tell you that the advice in this text has evolved over my several decades of experience, and has proven to be effective in my own career and in the lives of those I have mentored.

Honest advice from this grandfather-professor may not be given a high degree of relevance in your life. Obviously, you, the emerging generation of engineering academics, have the freedom listen to us "old guys" or not, and you have the opportunity to chart your own course. This leads to my bottom line summary piece of advice: *Think rationally and do your best to live your life with integrity, doing things that really matter* – by introspection, you should establish your own "compass" that will help you find a way around the obstacles you are likely to encounter.

Acknowledgements

Several colleagues at MIT, the University of Texas, Purdue University, and Texas A&M University have contributed information on recent promotions that led to the statistics of Tables 1 and 2. This text was greatly improved by including this information and I am pleased to express my appreciation to these un-named colleagues. I have not named the points of contact nor the departments to protect the privacy of the individuals whose data I included in the statistics.

I am delighted to recognize the collective contributions made by many good friends, colleagues and former students during my tenure at three fine universities:

- *the* University of Virginia

- Virginia Polytechnic Institute and State University

- Texas A&M University

In addition to the substantial advancement of my career and intellectual enrichment enabled by these collaborations, I have learned to enjoy several variants of the same institutional rivalry jokes, but told with differing signs!

I am especially indebted to the "Junkins Mafia" – my current and former students. You guys are truly a part of my extended family. You are also the main reason it is fun for me to wake up every morning and think about the day ahead. My claims on your career accomplishments are only in helping you set some good initial conditions, and maybe I have been of some marginal help in setting the sails for your career. I may have also contributed significantly in helping you develop an exceptional tolerance for pain (i.e., my humor!). You collectively possess a diverse set of technical competencies which simultaneously makes me jealous and fills me with joy. So now go, and make me even prouder of your accomplishments!

I am also pleased to warmly acknowledge the consistent love, support and numerous other contributions of my life partner and dear wife, Elouise. She, better than anyone else, appreciates the numerous investments we have made together along the wonderful path we have traveled together. She, more than anyone else, has brought vital measures of good judgment, stability and polish to assist me in virtually all aspects of my non-technical life. I could easily write extensively on this subject; cultivating a happy home is incredibly enhancing to achieving professional success and more importantly, to enjoying a great life. On the other hand, there is always a measure of fate and good luck involved, and in this regard, I am truly blessed to be well-married. Saying *thank you Elouise* is a hopelessly inadequate tribute, but you are the most intuitive person I know – you alone know what is in my heart.

I express my sincere thank you to my good friend, Clifford Fry, PhD; he proofed an early draft of this book and made constructive suggestions that very significantly improved the focus and style. My colleagues David Levinson and Yang Chen helped me eliminate grammar and typographical errors present in the first printing.

References

1. Written and oral confidential input from anonymous colleagues at MIT, Purdue, Texas A&M, and the University of Texas, (data and information enable me to compute the statistics in data in Tables 1 and 2).

2. http://en.wikipedia.org/wiki/Willard_Gibbs.

3. *von Kármán, T., Collected Works*, (4 Volumes), Von Kármán Institute, 1975, Rhode St. Genese, BE.

4. Filippone, A., *Advanced Topics in Aerodynamics, History Notes on Theodore von Kármán* (1881-1963), http://www.uady.mx/sitios/ingenier/fluidos/estudio/notas/historia%20y%20actualidad.pd

5. von Kármán, T; Edson, L., *The Wind and Beyond - Theodore von Kármán, Pioneer in Aviation and Pathfinder in Space*, Little Brown and Co, 1967, Boston, MA.

6. von Kármán, T., *From Low Speed Aerodynamics to Astronautics*, Pergamon Press, 1961, London, UK.

7. von Kármán, T., *Aerodynamics - Selected Topics in the Light of their Historical Development*, Cornell University Press, 1954, Ithaca, NY.

8. Hanle, P. A., *Bringing Aerodynamics to America*, The MIT Press, 1960, Cambridge, MA.

9. Gorn, M. H., *The Universal Man, Theodore von Kármán's Life in Aeronautics*, Smithsonian Institution Press, 1992 Washington, DC.

10. Committee on Science, Engineering, and Public Policy, *Rising Above the Gathering Storm*, National Academies Press, 2007. http://www.nap.edu/catalog.php?record_id=11463

About the Author

John L. Junkins, PhD holds the rank of Distinguished Professor of Aerospace Engineering at Texas A&M University. Dr. Junkins also holds the Royce E. Wisenbaker I Endowed Chair and is designated a Regents Professor. He has been a member of the Texas A&M faculty since 1985. He held prior academic appointments at the University of Virginia and Virginia Polytechnic Institute and State University. Prior to entering academe, he was an engineer at NASA and McDonnell Douglas Astronautics Company (now Boeing) during the Apollo program.

Dr. Junkins has expertise in spacecraft dynamical modeling, guidance, navigation, and control. His work spans theoretical, computational, experimental, design and invention aspects of advanced space systems technology, and his results have supported a dozen successful space missions. He has also invented a number of optical sensors and supports technology transitions to government and industry. He is the author of 6 engineering textbooks and several research monographs. He is also the author of over 230 archival papers and holds a number of patents. He is a prolific mentor, having directed over 100 graduate students including four dozen doctoral students and two dozen post-doctoral researchers. Over 20 of his technical offspring are professors, giving rise to 3 generations of PhD descendants.

Dr. Junkins is a Member of the National Academy of Engineering and is a Fellow or Honorary Fellow of several relevant professional societies. He has been honored as the recipient of over a dozen national and international professional awards in recognition of his research, scholarship, teaching, and mentoring. He has been honored as a distinguished alumnus of both of his alma maters, Auburn and UCLA. He proposed and is the founding interim Director of the Texas A&M University Institute of Advanced Study.

Personal/Family: Dr. Junkins grew up in Dalton, Georgia. While in high school, he was more interested in football and girls than academics, but following President Kennedy's inspirational speech, he found his calling. Upon graduating from Auburn University, he married Nancy Elouise (Click) in 1965. They have two adult children: Kathryn (Junkins) Sarpong, a DVM who conducts her veterinary practice as an owner/operator of Metro Paws animal hospital in Dallas, Texas. J. Stephen Junkins is a computer scientist at Intel in Bend, Oregon. Stephen has two children (Anna 15 and Orion 12), Kathryn has two children (Abigail 9 and Elouise 5). John and Nancy Elouise presently reside in College Station, Texas and enjoy frequent visits with family and friends at their weekend home on Stillhouse Hollow Lake, near Salado, Texas.